The External Environment

Proceedings of ITEST Workshop

October 26-28, 1990

Robert A. Brungs, S.J., Editor
ITEST Faith/Science Press
3601 Lindell Blvd
St. Louis, Missouri 63108

Published by:
ITEST Faith/Science Press
3601 Lindell Blvd.
St. Louis, Mo. 63108

ISBN 0-9625431-3-6

Printed in the United States
of America Versa Press

TABLE OF CONTENTS

FOREWORD

This is the middle volume of a trilogy on the human relation to creation and the relation of the created universe (including us, of course) to God. In the first Workshop ITEST considered "the inner environment." This was divided into three aspects: clinical investigation, health care delivery and the economics of health care. I personally found it interesting that the workshop focused on the just distribution of the fruits of what we have already accomplished rather than the pioneering work going on in the laboratory. This, I think, reflects a mentality different from that of five years ago when the accent would have been on the future. This may be a symptom of a significant cultural, and maybe religious, retreat from our once wide open preoccupation with the future. This should be of concern both to the scientific and religious communities.

This middle workshop centered on "the external environment," that is, on all of creation external to ourselves -- external to us, mind you, not separate from us. The third workshop will deal with Jewish and Christian perspectives on creation, on our part in it and on its relation to God, its creator and sustainer.

In this second workshop ITEST has endeavored to the best of its ability to learn the realistic extent of environmental problems. In any movement as broad as the environmental movement there will be a mixture of fact and fantasy, of reason and emotion, of ignorance and knowledge. In any movement in which advocacy plays a large, maybe even dominant, role, there will always be a combination of information, misinformation and disinformation. Furthermore, when an effort becomes fashionable and attracts celebrities as well as a great deal of media attention, slogans often replace thought. This meeting was held in the hope of being able to penetrate to the real dimensions of problems like the greenhouse effect, acid rain and energy considerations and their impact on individuals and local and global society.

The attempt honestly to separate fact and fancy determined the make-up of the panel of essayists. The panel finally assembled included a physicist, an atmospheric scientist and a biologist. We recognized that the human does not live on science itself so we added two "regulators" to the mix. But the human does not live on science and regulation, so we made the mix even more interesting by fortifying it with a lawyer and a political scientist. Of the scientists, one is an academician, one a consultant to the government and the third an industrial scientist. We tried to include essayists from the more "radical" wing of the environmental movement, particularly with regard to the greenhouse effect. I must admit that I was surprised by the stipends (an order of magnitude greater than we could even consider) demanded by the superstars of the greenhouse debate. In my naivete I thought that the seriousness of the issue would keep stipends within reach in the interest of informing people and mobilizing them to work for improvement. So much for that notion! Actually, this was a blessing in disguise because, looking closer to home, we were able to enlist the expertise and help of Ben Abell.

In many respects, somewhat to my dismay rather than to my surprise, the feeling of the discussion was quite similar to that at the workshop on the inner environment. There was a tone, a perception, that we cannot multiply technological achievement without serious thought to its long-term effects both on the environment and on ourselves. I find myself ambivalent about this mood. On the one hand I personally and ITEST corporately have promoted a search for meaning in the welter of scientific and technological advance. That is our main goal. On the other hand, that search has been conducted in a radical openness to advance. I hope I am wrong, but I am beginning to find a sense of disillusionment with science and technology in our society. I think some of that was present in this meeting.

There are voices in our society that seem not to want to set reasonable limits to our scientific and technological efforts but want to retreat from our present levels of development. I personally am concerned about that sense of retreat. Nonetheless, in the March 1991 workshop on "Christian and Jewish perspectives 'on the creation'" we shall, from a Christian view, have to cope with the "limits" of a redeemed world, the world in which God became incarnate and remains with us in the church. These "limits" are really opportunities in that they provide the true dimensions of the world in which the Kingdom of God is growing.

It seems as if we are seeing the beginnings of a sea change in our approach to significant issues arising from the accelerating advances in science and technology during the last thirty to fifty years. I would hesitate to make a firm prediction on the basis of only two meetings, but this is a phenomenon worth watching. In the famous words of one of our members "we'll know more later." Perhaps that phrase is an antidote to a loss of nerve either socially or religiously -- we'll always learn, if we care to do so.

I want to congratulate and thank our essayists for their fine work before and during this workshop. They contributed a great deal from their expertise in areas like climatology, energy, the Clean Air Act of 1990 and acid rain, law, political science, biotechnology and waste management. More, they contributed a great deal of themselves, of their humor and of their concern for the planet and its people. They deserve the thanks of all of us. As you will note as you read this volume, there are no interventions from Ms Nancy Kete of the Environmental Protection Agency. As coincidence would have it, Congress finally got around to debating the Clean Air Act of 1990 on the very weekend of the workshop. Ms Kete had to remain in Washington to testify in behalf of the Bill and we were consequently (and unfortunately) deprived of her input.

I would also like to thank the participants for their patience, their humor (much of which has been edited out of the Proceedings to save space) and their charity toward each other and toward me. Let me tell you why we must save space: we can bulk-mail these books only if they are less than a pound. So, in one sense, beyond 170 pages I have to edit by weight. Actually, this gives a tighter edit and one, I think, easier to read. I wish, also to thank the staff (Sister Marianne Postiglione, RSM and Sister Rosemarie Przybylowicz, OSF) for their dedicated behind-the-scenes work. This effort, never seen and rarely adverted to, is absolutely indispensable to a successful meeting. Finally, I would thank the ITEST Board of Directors for their continuing promotion of the work of the group.

Robert A. Brungs, S. J.
Director: ITEST

Human Influence On Climatic Change

Benjamin F. Abell

Benjamin F. Abell is a Professor of Meteorology/Mathematics at Parks College of Saint Louis University. He received a B.S. degree in Professional Meteorology and Statistics from Saint Louis University. Mr. Abell has worked for the National Weather Service at the Office of Climatology, the Office of Meteorological Research and the Analysis and Forecast Facility at Suitland, Maryland. He has been an on-air weather forecaster for a number of radio stations since 1972 and has been active as a meteorological consultant to Emerson Electric, The Olin Corporation, Mobil Oil, and the U.S. Air Force. He has designed courses in hydrology, air pollution and the meteorology of severe storms at Parks.

The climate of the earth has changed from probably the earliest geological ages, four and one-half billion years ago when the earth was either formed or severely metamorphosed, down to the present. This climate change will presumably continue for many ages yet to come. Both the paleoclimatic record and the more recent historical climatic record produce evidence of climate change.

Paleoclimatic evidence of climate variation is available from ice cores and marine sediments combined with radioactive dating techniques. Tree ring data, particularly from stress species such as the bristlecone pine, yield a continuous record of climate variation dating back 5000 to 8000 years. Paleoclimatic evidence becomes fragmentary and ultimately disappears for the oldest geological periods. Some evidence enables one to draw conclusions concerning extremes of climate over the last one million years. A relatively continuous paleoclimate record is available over the last 100,000 years. This record yields quantitative estimates of a number of climate variables.

Meteorological instrumentation has been available for more than 200 years. Synoptic records have been available since the last half of the nineteenth century. Prior to meteorological instrumentation, manuscripts record information regarding crop yields, drought, winter severity and variation of inland sea level.

Climatic Change

We live in an unusual epoch of earth's climate history. There is considerable ice in both polar regions. Although there is fragmentary evidence of extensive continental glaciation 600 million years ago and again 300 million years ago, the poles were ice free during most of the earth's history. The earth was considerably warmer 65 million years ago than it is today.

The last ice age was at a maximum 20,000 years ago. At that time, the North American ice sheet extended as far south as the Missouri River. Isostatic studies estimate that there was 5,000 feet of ice over what is now Toronto, Canada and 8,000 feet of ice over the Hudson Bay region. While the ice in both hemispheres has spectacularly retreated to its present limits, there have been several retreats and advances of the ice sheets since the last extensive glaciation. These variations have both responded to climate change and contributed to further climate change through feedback mechanisms.

Four epochal climate changes have occurred since the last ice age. The post glacial optimum, which peaked between 5000 and 3000 B.C., was a warm epoch. The extent of land ice and sea level were similar to today. Interestingly, the minimum land ice probably occurred around 1500 B.C. when the warm epoch was on the wane. Evidence of ancient plant and bog growth indicate warmer sea and land temperatures in high latitudes. Vegetative belts thrived at higher latitudes and altitudes. The European snow line was 300 meters above its lower limit today. The storm tracks and resulting precipitation belts shifted further poleward. Fluvial erosion in Antarctica dates to this period.

This was followed by a cold period which is now generally referred to as the post glacial climatic revertance. The lowest temperatures occurred between 900 and 450 B.C. There was a sudden reformation of the Arctic ice pack above 75 degrees North Latitude while the forest in Russia spread southward to the Dneiper River. Many of the present Rocky Mountain glaciers formed at this time.

The secondary climatic optimum from 1000 to 1200 resembled the post glacial optimum in many respects but was thought to be milder. There was no drift ice near the southeast and southwest coasts of Greenland around 1200 as Europeans established colonies on Greenland and Iceland. Vineyards in Europe were established 300 miles farther northward and there is evidence of warm droughty conditions in the mid and upper Mississippi Valley at this time.

The Northern Hemisphere cooled off again during the Little Ice Age (1430-1850). This cool period was not as severe in the Southern Hemisphere. A large expansion of the Arctic ice pack contributed to the failure of the colonies on Greenland. The relatively high levels of the Caspian Sea around 1800 was the result of many years of precipitation exceeding evaporation over the drainage basins which supply the Caspian. The storm tracks and precipitation belts migrate equatorward during cold epochs and poleward during warm epochs.

There have been several large and many smaller climate changes in the last one million years. Many of these changes occurred simultaneously over the entire earth at irregular intervals. These changes were of unequal duration and intensity. The climate will continue to change.

Natural Causes of Climatic Change

At least six mechanisms are capable of producing changes in the climate without human interference. They are variations in the eccentricity of the earth's orbit around the sun, the precession of the equinoxes, changes in the obliquity of the ecliptic, continental drift and mountain building, volcanic activity, and variations in solar output. The first three mechanisms are collectively referred to as the Milankovich Theory.

The eccentricity of the earth's orbit refers to the shape of the path which the earth describes as it orbits the sun. This eccentricity varies over a period of 95,000 years from nearly circular, as it is at present, to more elliptical then back to nearly circular again. At present, the earth is closer to the sun in January and farther away in July. The difference is about 3 percent of the mean distance between earth and sun, but it can be as great as 9 percent as the earth's orbit becomes more elliptical. Accordingly the variation in the amount of solar radiation received at the top of the atmosphere from January to July will vary from about 7 percent for low eccentricity to 20 percent for high eccentricity. This affects the severity and lengths of the seasons producing opposite effects in both hemispheres at any one time.

As the earth rotates around its axis, it slowly wobbles like a top. This wobble is called the precession of the equinoxes. At present, the earth is closest to the sun in January and farthest away in July. In 11,000 years, this will be reversed and the earth will be closest to the sun in July. After 22,000 years, the earth will once again be closest to the sun in January. Like the variation in the eccentricity of the earth's orbit, the precession of the equinoxes will affect the severity of the seasons producing opposite effects in each hemisphere.

The axis of rotation of the earth tilts at about 23 1/2 degrees away from a line drawn normal to the plane of revolution described by the earth's path around the sun. In other words, the earth tilts toward or away from the sun at 23 1/2 degrees (the obliquity of the ecliptic). This tilt is not constant but varies from about 23 3/4 degrees to 24 1/4 degrees and back again over a 42,000 year period. A small tilt would produce less seasonal temperature variation than a larger tilt. This would mean milder winters and cooler summers. Despite milder winters, the smaller tilt could encourage glaciation due to higher amounts of precipitation brought about by higher moisture content in milder winter air. Then an interesting positive feedback mechanism could activate, as increased glaciation raises the earth's albedo (amount of incoming solar radiation returned to space by reflection and back scatter). This would promote further cooling.

The current trend of the three mechanisms comprising the Milankovich Theory points toward a cooler earth with increased glaciation and perhaps a new ice age.

The continents are moving - very slowly measured over our average life span. Sections of the Pacific Ocean are spreading apart about 5 cm each year. Spreading rates in the Atlantic and Indian Oceans are less than half that figure. In geologic time, the movement is impressive. The continents were lumped together in a single super continent 200 million years ago. Africa and South America were joined as were North America, Greenland and the Eurasian land mass. The Atlantic Ocean is geologically new.

Not only have the continents changed position, but ocean size varies. Thermal contrast between land and water controls the world's gargantuan monsoonal circulation and influences the position of the mid-latitude jet stream and storm track. Both the monsoonal circulations and jet positions change from year to year and over historic and geologic time.

Erupting lavas and diastrophic movements build mountains, while weathering tends to reduce land areas to a single level. This would not only impact temperatures but would control precipitation and a lack thereof over large regions of the earth. This in turn would influence global climate.

Volcanic eruptions spew both gases and pulverized fine pyroclastic debris into the atmosphere. Volcanic dust can extend into the stratosphere following particularly violent eruptions. Once in the stratosphere, these dust particles may remain for a year or two and back scatter some incoming solar radiation. This cools the earth. Cool weather followed the gigantic eruptions of Asama in 1783, Tambora in 1815 (New England's year without a summer), Krakatoa in 1883, and Katmai in 1912. However, the cooling is only evidenced for several years following the eruptions. It would take widespread volcanic activity over many years to produce climate change for a longer period.

Finally, variable solar output could produce a change in climate. Our sun is a variable star. The solar constant is not constant. Of course, it changes due to external (external to the sun's output) influences such as the changes in the eccentricity of the earth's orbit and the precession of the equinoxes. Moreover, variations in both short wave solar radiation and corpuscular radiation from the sun vary over periods of years in cyclic and quasi-cyclic manners. Perhaps the 11-year sunspot cycle is most familiar, but there is evidence of both a 22-year cycle and several 90-year cycles in the last two and one-half centuries. There was little evidence of sunspots from 1645 to 1715. This period of little or no sunspot activity has been labeled the Maunder minimum.

The earth was relatively warm during sunspot maxima in the twelfth and thirteenth centuries. Interestingly, the hot droughty years in the central United States in the 1930's and 1950's coincided with the increasing curve of the second 11 year sunspot cycle within the sometimes elusive 22-year cycle. For years, investigators have been attempting to statistically link climate events with solar cycles. There appears to be something there but it is a very elusive something. Moreover, statistical correlation does not prove physical causality.

During the active phase of the solar cycle, the number of sunspots is relatively high. Sunspots are magnetic disturbances on the surface of the sun. They are cooler than the surrounding solar photosphere. During the active sun phase, solar flares and prominences become frequent. There is a measured increase in solar radiation in the ultraviolet portion of the spectrum eight minutes following a solar flare. This is followed, 18 hours later, by an increase in the solar wind (corpuscular radiation). This increase in corpuscular radiation in turn triggers disturbances in the earth's magnetic field and is often responsible for brilliant auroral displays.

Temperature, density, and ionization increase in the heterosphere, high above the earth's surface during the active sun phase. In fact, the atmosphere expands and bulges outward. The picture becomes murky when a cause and effect association is attempted between changes in density, heat and chemistry in the upper atmosphere and responses in the complex thermal and pressure fields in the lower 50 km. of the atmosphere. It appears that longer term warming is associated with an active sun and cooling with a quiet sun.

In addition, the standard theory holds that the sun has increased its luminosity at a fairly constant rate over the last four and one-half billion years. Periodically, this steady increase may be interrupted by abrupt changes in luminosity amounting to as much as 20 percent. The calculated interval between these abrupt decreases in surface luminosity is 300 million years, which corresponds to the time scale of the epochs of glaciation on the earth.

There could be many primary causes of climate change. The six major natural causes, reviewed in the preceding sections, could trigger feedback mechanisms which could either accelerate or dampen the change. An example of climatic feedback is increased glaciation which then increases the earth's albedo and further lowers the earth's temperature stimulating additional glaciation.

The Human Factor

Most human climate modification is subtle and difficult to evaluate against the background of natural climatic fluctuation. Local climate modification by human activity can be dramatic. Some examples would be increasing temperatures caused by urbanization and industrialization and regional changes brought about by soil mismanagement or destruction of a natural resource such as forest or grassland. Evaluating human influence on large scale climatic fluctuation is an entirely different matter. The four human activities capable of producing climate change, which are most often cited by present day investigators, are the expansion of the urban heat island, increasing the atmospheric greenhouse effect, increasing the back scatter of solar energy due to particulate pollution, and deforestation and desertification.

Urban industrial centers often create a local climate which differs from that of the surrounding countryside. This is called the heat island effect. A city retains more of the day's heat than the surrounding suburban and rural areas. The nighttime temperature contrast between city and country is greatest under clear skies and low wind. This difference disappears under overcast skies and/or windy conditions. Sensible heat is also added to the atmosphere in an urban-industrial complex due to energy consumption and industrial processes. Local climates are spectacularly and irreversibly changed by increased urbanization, but the contribution to global climate change is miniscule.

In the late 1800's, the average temperature of the Northern Hemisphere began to rise. This temperature rise continued through the first half of the twentieth century and was more noticeable in mid and high latitudes than in tropical regions. The generally accepted magnitude of temperature rise ranges from 0.5° C to 1.0° C. Scientists began taking a long hard look at this trend in the 1950's and many of them settled on the well established 15 percent increase in the atmospheric carbon dioxide content as the primary cause. Carbon dioxide is transparent to short wave solar radiation but it selectively absorbs long wave terrestrial radiation. This increases the atmospheric greenhouse effect and should produce a rise in the mean temperature of the earth.

The world climate refused to cooperate with this theory and began cooling around 1955. The cooling trend accelerated in the 1960's and made up about half the ground lost during the earlier warming period. The carbon dioxide greenhouse theory began to fall into limbo. Some scientists sought to explain the cooling in terms of increased back scatter of solar radiation due to an increase in particulate pollution. The cooling trend leveled off in the 1970's. The temperature trend has shown considerable year to year and region to region fluctuation since 1975, but the overall trend appears to be one of renewed warming. As a result the carbon dioxide greenhouse theory is experiencing a resurgence.

What is the bottom line? Atmospheric carbon dioxide content has increased about 17 percent over the last 100 years. Scientists are now turning their attention to increases in atmospheric trace gases such as methane, ammonia and several of the chloroflurocarbons which may also increase the atmospheric greenhouse effect. These conclusions appear theoretically sound, but there are some serious problem areas with this theory.

Water vapor, which is extremely variable from one geographical location to another and from season to season within many geographical areas, is also transparent to solar radiation and selectively absorbs terrestrial radiation. The water vapor content of the atmosphere is several orders of magnitude greater than that of carbon dioxide, and the carbon dioxide content is much greater than that of the combined aforementioned trace gases. If water vapor content changes, its greenhouse effect would dwarf that of carbon dioxide and the trace gases.

Is the water vapor content of the atmosphere changing? It is difficult to say. If an increase in carbon dioxide triggers a small temperature rise, one could reason that the resulting temperature increase would promote more evaporation and, therefore, more water vapor. This could contribute to runaway warming. However, an increase in atmospheric water vapor should lead to more cloudiness and precipitation. This would increase the earth's albedo and lead to global cooling. Perhaps the increase in atmospheric carbon dioxide is not the monster that many of the modern prophets of doom say it is.

Combustion due to human activity contributes to climatic change in another manner by adding aerosols to the atmosphere which increase the earth's albedo. Although most atmospheric particulate matter is the result of natural processes such as weathering, the human contribution has an impact. While dark aerosols such as soot may actually absorb more solar radiation, light colored particles reflect solar energy. Both dark and light particles back scatter incoming solar radiation. The overall effect appears to be an increase in the earth's albedo. Aerosols may also absorb some terrestrial radiation, but the cooling due to the albedo increase should predominate.

Finally, human modification of the earth's vegetative cover can have a dramatic impact on microclimates and regional climatology as well as some input on global climate. The greatest impact is realized through desertification and deforestation. Deserts continue to spread. This often results from overgrazing mainly carried out by practices of nomadic cultures in the Third World. When overgrazing is combined with periodic drought, the desert migration into semiarid regions accelerates. The tragic events in the African Sahel over the last 10 to 20 years is a dramatic illustration of desertification. Left alone for many years, an area could rebound. Sadly, most desertification is irreversible.

Modern societies clear forested areas in order to increase agriculture and logging, drain swamps, stimulate herding and improve chances in war. This often ruins a landscape over large regions. The most widely used and effective tool in the human arsenal for land clearing has always been and still is fire.

Wildfire due to lightning is one of the oldest of natural phenomena. Many ecosystems not only tolerate but encourage fire. It is the dominant selective force for determining the relative distribution of certain species, and it stimulates and actually feeds effective nutrient recycling. Humans adopted fire as a land clearing tool thousands of years ago. Originally, people brought their fire to North America across the land bridge from

Asia. The next wave of fire practice came to the New World from Europe.

When Europeans crossed the Appalachian uplands to settle in what is now the central United States, they encountered an immense sea of grass, the prairie. They should have seen vast forests. Climate will support widespread grasses in only two regimens. One is located in the broad belt between the great low latitude deserts and the equatorial rain belt or in a monsoon regime, where a lengthy dry season promotes tropical savannah growth. The other regions which can naturally support widespread grasses, are the mid-latitude semi-arid areas produced by rain shadows in the lee of mountains. Sparse precipitation in these areas produce short grasses and scrub.

The prairie was the result of annual late summer and fall fire setting by the hunting and gathering native Americans. Europeans continued the practice in order to promote agriculture. Forests have made a comeback in some areas where Europeans replaced native Americans. This is an example of human modification of vegetative cover on a grand scale. The regional climatology responds to the changing albedos brought about by a new vegetative cover.

Both desertification and deforestation increase the earth's albedo over large areas and this immediately changes the regional climate and eventually world climate by altering the location and intensity of precipitation belts.

Societal influence on regional climate change is profound and in many situations irreversible. Societal influence on global climate change is another question. The impact is simply not known. Certainly, immediate attention and corrective measures should be brought to bear on desertification and deforestation, but there is no evidence of runaway global warming due to increasing carbon dioxide content in the atmosphere. Perhaps the combination of increasing carbon dioxide content and particulate pollution is resulting in a climatic standoff.

Caring for the Environment: It Begins with Each of Us

Dan Bennett

Mr. Bennett, an environmental specialist for the Missouri Department of Natural Resources, was a former trainer and field manager for the Missouri Coalition for the Environment. He was the local coordinator for the National Campaign Against Toxic Hazards in 1985. Duties at the MDNR include: factory inspections evaluating hazardous waste compliance; speaking on recycling, household hazardous wastes and other issues; working with other agencies and community leaders on waste minimization techniques and programs to improve environmental quality, preparing site investigations for Missouri sites proposed for the Federal Superfund. His memberships include: the St. Charles Solid Waste Committee, the Air and Waste Management Assoc., the Missouri Waste Control Coalition, the Gateway Chapter of Certified Hazardous Materials Managers and the St. Louis Environmental Education Network.

The twentieth anniversary of Earth Day was heralded by an almost unprecedented response from the media, politicians, veteran environmentalists and other grandstanders, but the most important impact may have been on the many people who have heard of the many issues, and only now have decided to get involved.

Many of the problems we face have developed since the Industrial Revolution, and it is the factories and power plants that frequently are singled out as the root of most problems. But human culture has always been in a dynamic, yet precarious, balance with nature. The early hunter and gathering tribes were forced to migrate in order to avoid long term degradation of their environment. The wandering herd, on which their livelihood was based, would move on as their food supply diminished and the humans would follow. Each member of the tribe followed established routines that prevented depletion of the herds, produced little waste, and kept them all in a natural cycle that could replenish itself as the herds and tribes moved on. As the early civilizations found stability through agriculture, other human industries developed such as mining and metallurgy, medicine and pharmacology, building and architecture and environmental challenges increased. Many of the earliest civilizations did not meet those challenges. Poor sanitation or the lack of soil or water conservation forced many cultures to move on, leaving the problems behind. The hunters and gatherers had learned that there were ways to adapt with nature to assure a stable life-style. The challenge to our ever-changing industrial society is for each of us to do the same.

Archaeologists studying ancient cultures find a wealth of information at each dig. Ancient American Indian skeletons are found adjacent to ancient campfires, stoneworking fossils and garbage. These people lived and died and dumped all in the same area. Environmental complications were not obvious since they would move on allowing decomposition and other biological activity to occur in their absence. Early town dwellers lived in essentially the same way. Horticultural tools and pottery showing the imprint of many kernels of grain identify the earliest agricultural regions. But as the human population in the isolated centers of civilization grew, the problems became more obvious. The lack of adequate sewage disposal was a major contributor to the ravages of epidemics in Middle Age Europe. It wasn't until 1842 that disease was linked to environmental conditions, and even then garbage and sewage disposal in some major cities consisted of piling the refuse in the streets. In 1894, Harper's Weekly reported that "the garbage problem is the one question of sanitation that is uppermost in the minds of local authorities." Yet, open garbage dumps in America persisted into the 1960's. Even though city dwellers learned to remove sewage and garbage from their city limits, they were slow to realize the environmental impact outside of their boundaries.

The City of St. Louis found the Mississippi River a convenient outlet for sewage wastes, but for centuries ignored the environmental impact downstream. Primary treatment of sewage removed solids from the discharge, but millions of gallons of water saturated with waste were still dumped, promoting pathological organisms and disrupting ecological cycles downstream. The city is only now completing a secondary treatment facility that will remove 90-95 percent of the total sewage being discharged.

Early mistakes were made in garbage disposal, as well. St. Louis used wetlands for disposal of much of its garbage for years. Now there is an extensive area along the river bank, north of downtown, where groundwater movement parallel to the river is transporting a host of contaminants underneath the fill. This area is especially complex, since historical mismanagement of hazardous materials has contributed significant physical and chemical properties to the groundwater. Designating a responsible party to clean up that contamination is often difficult, since it is questionable if a waste was disposed of in place, or has migrated from an upcurrent facility, or was dumped with municipal garbage years before.

St. Louis is not alone in facing such a conundrum. The EPA has identified a National Priority List under the Superfund and over 40 percent of those sites are former landfills. This is understandable, since it was not until 1965 that the Solid Waste Disposal Act (SWDA) provided guidelines for operating landfills and 1970 that the Resource Conservation and Recovery Act (RCRA) identified hazardous wastes and required alternative means for their disposal. Even so, landfills continue to receive hazardous materials. Many people overlook the fact that hazardous wastes are residues from hazardous materials, which frequently are included in consumer products. When unused products are disposed of they add their hazardous characteristics to the landfill, chemically reacting with other wastes to produce a toxic "soup" that migrates through the landfill as leachate. Factories that produce less than 100 kilograms (220 lbs) of hazardous wastes per month are unregulated by RCRA, and may not dispose of their wastes in a responsible manner. Also, more than 1200 regulated generators have been identified in the St. Louis area, with only four (4) state inspectors to evaluate the safety of their operations. Complaints of disposal of hazardous wastes into the trash, sewers and environment come in a never ending stream into the St. Louis office.

The EPA and other regulatory agencies are aware of the exemptions, missed details and loopholes that exist and are providing a constant deluge of new regulations intended to fill the gaps. But the new regulations often contain new contradictions and loopholes that serve to confuse waste generators and even overburden the regulatory agencies themselves. Currently, Missouri officials are considering legislation to support "waste minimization" that will provide industry with procedures and technology intended to reduce or even eliminate some hazardous and solid wastes.

The actions of government, industry and environmentalists have shadowed human activity throughout history and have drastically accelerated in the last twenty years, but we still cannot come to terms with the problems. Where should be the focus of responsibility? On industry, who wastes too much? On government, who reacts too little? Or on each of us, who as consumers or constituents the others are only trying to serve? There is no industry whose intent is to create hazardous waste or to degrade the environment . But in supplying consumers wants and needs, impact on natural resources and the environment is inevitable. Government response to environmental problems is often short-lived or misguided, and requires sustained involvement to produce results. The enormous public response to the recent Earth Day is a very good beginning, but no substantial results will be seen unless we all realize that it is only a beginning.

Public awareness is always the foundation of industrial and government actions. But that foundation is too often based on fear. When public opinion drives government into action, it is often in relation to a disaster, and the bureaucrats and politicians make a well-publicized, and often excessive, response. More knowledge on the part of the public may help to decrease the number of disasters and more effectively direct the regulatory response. The continuing education of the public or of the consumer may pursue two lines: communication with knowledgeable individuals and development of technical expertise. The first is more important, at least in the short-term. Those concerned about an environmental issue can broaden their perspective by writing, calling or visiting public officials, industry representatives, citizen-action groups, or educators. Since each individual source probably has a unique angle that is not immediately apparent, look for other sources of information. Books and magazine articles may give a general background that can help to interpret or substantiate the individual's contacts.

Now, it's not as though every newspaper headline should lead to personal investigative reporting, but each phone call will broaden one's understanding of the complexities of environmental issues. Some contacts will seem more communicative or down to earth than others. These people will usually be willing to offer specific references to broaden your technical capacity. It does not take a degree in chemistry to understand most environmental issues. However, most Americans do not understand the chemistry described on a box of laundry detergent. The dearth of students in the sciences has been described as a prime reason for America's flagging industrial output; but that lack of knowledge is perhaps the major stumbling block in society's pursuit of environmental solutions. Adults may not want to pursue organic chemistry as continuing education, but environmental studies can serve to enliven the hard sciences for youngsters. Having more people who are aware of chemistry and biology is crucial to understanding the challenges that will inevitably confront us in the future.

While awareness is the foundation of solving environmental issues and education is the framework, there is much more that individuals must do to have an impact on environmental issues. The key phrase is to "get involved." The St. Louis area has many local issues that will require public input to resolve. The city itself sends almost all of its trash into Illinois for disposal. With disposal costs always climbing, how can the city plan to insure continued disposal? The airport frequently makes headlines with the complicated disposal of radioactive waste there, and a major environmental cleanup at Weldon Springs in St. Charles County continuously provides issues that impact on the public. A recurring response to all of these issues is that the public would prefer the processing, storage or disposal of the variety of wastes to occur somewhere else. But where? Continued long term involvement of an educated public is essential to the political and technical issues involved. This will also involve state wide, national and international cooperation. Acid rain, destruction of the ozone layer, global deforestation, and the greenhouse effect will have an impact that does not recognize national boundaries or local regulation. American consumer habits are a large part of all of these issues, and there will be no decisive action without changes in individual life-styles and American governmental leadership. Recommendations from industry or bureaucratic experts will never have the impact of an informed citizenship.

Humankind has always been dependent on its relationship with the environment, and has always been dependent on how we interact culturally to define that relationship. In order to deal with future challenges we must learn what each must do to correct the mistakes of the past. Unlike the fallen civilizations of the past, we can no longer walk away from the challenges.

Biotechnology and the Environment

James F. Kane

Dr. Kane received his B.S. in Biology in 1964 from St. Joseph's College in Philadelphia, PA., and his Ph.D. in 1969 from the State University of New York, Buffalo. From 1968 to 1970 Kane was a Postdoctoral Fellow in microbiology, Baylor College of Medicine. He was an assistant professor (1970), associate professor (1975) and vice chairman (1980) in the Department of Microbiology and Immunology at the University of Tennessee Center for Health Sciences. Kane became Senior Scientist/ Section Head for Restriction Enzymes at Bethesda Research Laboratories in Gaithersburg, MD in 1981. In 1982 he joined Monsanto as a Senior Research Specialist in the Biocatalysis group of Corporate Research Laboratories. In 1984 Kane initiated studies on the microbial production of bovine somatotropin, and was appointed Senior Group Leader of Animal Sciences Division in that same year. In 1986 he assumed the additional responsibilities for fermentation process development. Kane was appointed a Fellow in the Monsanto Fellow Program in 1987.

I don't think I have ever had to present a 15 minute talk on a topic so broad before! I would like to begin by discussing each of these broad topics individually, and conclude by tying them together in a hopefully coherent fashion.

The word environment means the combination of external or extrinsic physical conditions that affect and influence the growth and development of organisms. What the specific external or extrinsic physical conditions are depends on your personal biases and areas of interest.

In this workshop, we examine the environment from the perspective of the effects of our activities on the climate, the accumulation of waste, and energy production and utilization. In my presentation, the role of an emerging technology on these and other environmental considerations will be discussed.

It is clear that the environment is a major consideration in the mind of the public. Press releases and news coverage literally bombard the public with the dangers of pesticides and agricultural chemicals on our food and water, the dilemma of disposing of solid wastes as our landfills begin to reach capacity, the problems of acid rain on the defoliation of our forests, the greenhouse effect influencing the temperature of the earth, the destruction of our rain forests and the concomitant extinction of various life forms, and the disposal of nuclear waste. The list of environmental concerns seems to grow almost daily as we read of yet other problems that impact or will impact us as we conduct our daily lives.

It is also clear that the term *Biotechnology* elicits some specific impressions. In point of fact, biotechnology is simply the use of a biological system or catalyst in a process used to make specific products for the consumer. Biotechnology is a very old technology dating back to early Egyptian cultures which used the well known process of fermentation to make alcoholic beverages. In addition biological processes are used in the manufacture of foods, cosmetics, beverages, vitamins, sweeteners, animal and plant breeding and waste (sewage treatment) management. When you see a product on the shelf that contains the statement "All natural ingredients" several of the components of these products have been made using biotechnology. Why then does the term biotechnology elicit such fear? What has changed in the minds of the public? I think that the public has the perception that biotechnology is equivalent to *Genetic Engineering* which has the connotation of influencing the very structure of life itself. But genetic engineering is not new either. For years, animal and plant breeders have used genetics to increase the productivity, longevity, stability, and quality of plants and animals. What has changed is that we (scientists) have discovered through research a new technique or method of doing genetic engineering. So much has been written in the press about this new method that this method has become synonymous with the term genetic engineering. While genetic engineering is a much broader area of scientific research, I am going to use the term genetic engineering to refer to this specific technique. In order to understand this new technique of genetic engineering one must understand a few terms.

The first term is *Deoxyribonucleic Acid*, better known as DNA. DNA is in effect a blueprint of life for the cell or organism. As the construction team assembles a building from a blueprint, the cells assemble all of the materials they need from DNA. DNA makes a cell what it is, and how it functions. We need to examine DNA more closely because of the central role this molecule plays in the area of genetic engineering. DNA is a polymer that is both simple and complex. Its simplicity resides in the fact that this very large molecule is composed of only 6 components. Four of these components are called bases, and have the chemical names of adenine (A), thymine (T), guanine (G) and cytosine (C). The other two components, a carbohydrate (sugar) and phosphate, serve to link the bases together. The molecule is composed of 2 strands bound together in such a way that A is opposite T and G is opposite C. The complexity of the molecule resides in the mechanisms that the cells use to transfer the blueprint information from the DNA into the proper components that make up the living organism. It is important to remember that the specific sequence of bases in the DNA is unique to each living organism. Thus, each cell contains a blueprint of life that dictates what products will be made by the cell.

A second term is *Restriction Enzymes*. These enzymes were discovered in the 1950's but their significance was not appreciated until the 1970's. One can view these enzymes as "scissors" that have the ability to cut DNA at very specific places regardless of the source of DNA. As a result the scientist can remove a very specific piece of DNA from one organism, and move it into another organism thus conferring upon the recipient a new property. An excellent example of this is found with the product Human Growth Hormone or HGH. HGH is given to people suffering from dwarfism or a dysfunction in their system that results in an insufficient amount of HGH being produced by their body. In the past the only source of HGH was the pituitary glands of cadavers, and as result there was not an abundant supply of this protein available to treat all of the people suffering from this disease. Scientists used restriction enzymes to "cut out" the HGH gene from DNA extracted from the pituitary gland of a cadaver. This gene was transferred to a bacterium called *Escherichia coli,* and now there is an unlimited supply of HGH using *E. coli* as the source of this protein instead of cadavers.

Now let us put this genetic engineering in the perspective of environmental questions. First of all, this technique is not a cure-all for environmental problems as some protagonists would have you believe. It is pure hype to propose that genetic engineering will bring an end to using chemicals in agriculture, and cure all environmental problems. Neither is it the beginning of environmental havoc, nor is it a path that leads away from sustainable agriculture as some antagonists would have you believe. Unfortunately, the antagonists desire an "absolute guarantee of safety," and no one, not even the most avid advocate of this technology or any other for that matter, can offer that. Why, because there are no absolute guarantees of safety in anything we do. We need to realize that genetic engineering is simply a technique that could have a major impact on some of the environmental problems that we now face.

For the remainder of this presentation let's take a view of the various types of genetic engineering experiments that could impact the environment. I will break these down into two areas; namely, genetically engineered plants and genetically engineered microorganisms.

Genetically Engineered Plants

We now have the technology to introduce genes into various plant species. This offers a number of unique opportunities some of which could have significant positive effects on the environment. Let us discuss a few of these.

1. Genetically engineering plants for increased resistance to diseases caused by insects and viruses. At the present time, diseases caused by insects and viruses are treated with chemical sprays or applications of insecticidal proteins that specifically destroy certain plant pests. Of these two possibilities the insecticidal proteins are the least threat to the environment. However, if one were to add insecticidal or anti-viral genes to the plants, then one would be able to reduce the amount or frequency of these chemical applications by the farmer, thus reducing the insult to the environment. The technology exists to confer upon some plants

increased resistance to disease. This is chiefly due to the fact that single genes can be added to the plant to bring about the desired effects. For example, a specific bacterium makes a protein that kills certain worms that feed on cotton plants. Currently, this protein is purchased by the farmer and dispersed by topical application across the fields. For this to be effective, the worm must ingest this protein as the worm crawls through the soil. This means that the bulk of added protein is wasted. A much more effective means of control is to place the protein at the site of feeding, namely, in the plant. This alleviates the need to spread this protein across the soil and reduces the amount of pesticide that is required. With this new technique of genetic engineering, the bacterial gene can be inserted into a cotton plant thus providing the plant with the capacity to make the bacterial protein. When the worm feeds upon the plant, it ingests the insecticidal protein and dies before damaging the plant.

An even better case can be made when the insecticidal gene placed in the plant replaces chemical sprays. Chemical sprays or applications while effective are nevertheless wasteful because the bulk of the chemical that is sprayed over the crop is never touched by the insect. By placing the insecticidal agent in the plant, environmental insults brought about by chemical spraying are obviated.

The positive effects of this technology seem obvious. The negative effects are less obvious, and reside principally in the minds of the antagonists who propose a limitless array of "what if" possibilities with few, if any, constructive suggestions.

2. Genetically engineering plants for increased resistance to herbicides. Most of the farmers use herbicides to reduce the amount of weeds in their fields thereby increasing the yield of their crop. These applications make good economical sense. A number of these herbicides however, are among the nation's worst groundwater pollutants. What is the solution? Well, one possibility is to switch from such environmentally unfriendly chemicals to chemicals that will kill the weeds, but remain environmentally friendly. What are the properties of an environmentally friendly herbicide? First, it would be effective in killing the weeds. Second, it would not enter the ground water but would be bound by the soil particles and remain at the site of application. Third, while remaining at the site it would be destroyed by the indigenous microorganisms in the soil to innocuous by-products in a relatively short period of time, say 2 to

3 weeks. Such herbicides do exist. One problem is that some of these herbicides are preemergent herbicides which are not specific to weeds. That is, these herbicides kill all types of grasses and flowering plants, and therefore must be applied before planting. If plants were engineered to be resistant to these environmentally friendly herbicides, then these herbicides could be applied after the plants have germinated. The advantage of such an approach is that you replace environmentally unfriendly chemicals with chemicals that do little or no harm to the environment while at the same time helping the farmer maintain optimum crop yields. Antagonists counter this with the comment "Isn't the only advantage to the Chemical Companies that make herbicides?" The answer is no. One must keep in mind that we can not solve all of the problems at one time. Basic research is required to understand how to optimize agricultural yields without the use of chemicals. In the interim, any approach which helps reduce the use of environmentally unfriendly chemicals should be encouraged.

3. <u>Genetically engineering plants for increased growth rates.</u> A novel approach to reducing herbicide applications is to make plants that grow at significantly faster rates such that the growth of weeds does not adversely impact crop yield. This is a much more difficult research project that involves more than single genes. We are talking about multiple gene clusters that have to be changed to reach this goal. A side benefit to this is that plants are the major utilizers of carbon dioxide in the atmosphere. An increase in the growth rate means reducing the levels of carbon dioxide and increasing the levels of oxygen in the atmosphere. Both of these effects significantly improve the environment.

Genetically Engineered Microorganisms

The idea of genetically engineering microorganisms (GEMS) engenders the most fear among the antagonists of this technology and the public as well. There is a fear that once released there is no way to stop or remove the microorganisms from multiplying and creating more havoc than these creatures solved. I think there is a general agreement that each potential opportunity should be evaluated independently. We do not have sufficient knowledge to make broad generalizations about the positive or negative effects of GEMs in the environment. There are, however, many opportunities for GEMs to have a positive environmental impact.

1. <u>Genetically engineer microbes to degrade toxic chemicals prior to their release into the environment.</u> My perception is that the word chemical is synonymous with *toxic chemical* in the mind of the public. People fail to realize the very positive effects chemicals have had, and continue to have, upon their lives. They are, however, acutely aware of the negative effects that toxic chemicals have. In the view of the public, toxic chemicals are bad regardless of levels. In response to this growing public concern, socially responsible industries are trying to reduce the amount of toxic waste materials that flow into the environment from various production facilities. Interestingly, it is this very fear of toxic chemicals that may help in the public acceptance of genetic engineering and biotechnologies dependent upon these techniques. Wouldn't it be wonderful if we could engineer microorganisms that would degrade these toxic materials to less toxic or perhaps useful chemicals, and help prevent the accumulation of these pollutants in the environment? While not an easy task, certainly some toxic chemicals can be treated with this technology. Mother Nature is incredible in its array of microbial populations and metabolic capabilities. Ideally, the GEM would be used in a treatment pond to degrade or reduce the level of toxic effluents emanating from the production facility. While generally confined to the industrial waste treatment site, the current fear is what will happen if this altered microorganism is released from the waste treatment site. Such microorganisms are not yet available; so to propose horrific outcomes is premature. It may be possible to make the microorganisms but these toxic chemical eaters may not work in the desired external environments, thus frustrating our attempts to solve the problem. In any case it is very important to understand the positives and negatives of the specific application before becoming fearful of the consequences.

2. <u>Genetically engineer microbes to degrade toxic chemicals already present in the environment.</u> At the present time, industries are attempting to cope with the presence of toxic chemicals in the environment. While this technology offers hope that some solutions can be found, it is a formidable problem to produce such microorganisms. In this case, the fears expressed above are exacerbated because the GEMs are being directly released into the environment. The typical "what if" scenario is that some type of SUPERBUG would be created that would create a worse situation than the one being corrected. Regulations are being proposed and should help in establishing the necessary safeguards to avert a disaster. We are not, however, infallible and mistakes will be made. The best we can do is to test the hypothesis and make a best

judgment after carefully weighing the risks and benefits.

3. <u>Genetically engineer microbes to become biocides.</u> As mentioned above, chemicals are currently employed to kill plant pests. One approach to reduce the load of chemicals in our environment is to use biological biocides. That is, use microorganisms to control the plant pests. This involves taking a microorganism that normally lives on the plant and adding some genes to this microorganism that makes it toxic to the plant pests. Of necessity, this involves the release of GEMs into the environment and brings with it the same fears described above. Namely, how do we control the GEM, and can we trust the scientists to know what the effects will be on the environment?

Summary

I would like to summarize this discussion by reminding you that this new technique called genetic engineering is not a panacea for all of our ills. Rather, it is a very powerful tool that offers us some real opportunities to have a positive impact on the environment in which we live. As research continues in molecular biology, biochemistry, microbial ecology, and molecular genetics we will begin to uncover the truths of nature and be able to use them to enhance the quality of our environment. I think it is premature to resist this technology. While I am not supporting an unbridled approach to the use of genetically engineered plants and microorganisms, I am suggesting a pursuit of this aspect of science to increase our understanding of the problems and potential solutions. I will close with a very appropriate quote attributed to Louis D. Brandeis: "More erroneous conclusions are due to the lack of information than to errors of judgment."

A Presentation on Economics and the Environment

Nancy Kete

Ms. Nancy Kete is on the staff of the Environmental Protection Agency in Washington, D.C. Because the Congress finally was debating the Clean Air Act of 1990 on the same weekend as this ITEST Workshop, Ms Kete was not able to be present. We are, however, grateful to be able to present her essay in these Proceedings.

Thank you very much for inviting me here to join you in a discussion of economic aspects of environmental solutions. Although I will discuss extensively one particular environmental solution that I am very familiar with -- the acid rain control amendment to the Clean Air Act and the way it incorporates the power of the market in solving the acid rain problem -- I would first like to frame that particular response in its broader political, economic, environmental and ethical context. Noting that the theme of this workshop is to explore the present and near future facts on how we treat the environment, it seems that the acid rain example is only valuable if described in its broader context.

The relationship between the environment and the economy will be one of the most explored themes in the 1990's. To my reading there are three, more or less distinct, ideas in people's writing on the environment/economy theme. The first idea is the broadest and it questions nothing short of our value system, including <u>whether</u> non-human life forms and non-market amenities such as a wilderness have intrinsic value, or whether we value such things only as commodities or providers of goods and services as a natural ecosystem is valued for providing timber and mineral resources. The second idea has to do with <u>how</u> we place value on, or <u>evaluate</u> what these traditionally non-market things add to our economy and how misuse or destruction of those non-market things both affects our well-being and results from our conventional notions of the economy. Once we have evaluated the contributions that market and non-market goods and services make to our well-being, the third idea questions what is the best way to enhance and protect the use and or preservation of those things we value so as to maximize that value.

Recognizing Environmental Values

I have the least to present formally on the first idea, although I highly recommend the 1989 book by Robert C. Paehkle, <u>Environmentalism</u> <u>and</u> <u>the</u> <u>Future</u> <u>of</u> <u>Progressive</u> <u>Politics</u> for, among reasons, its excellent history of what is now termed "environmentalism" in North America. As religious leaders you certainly must recognize the Judeo-Christian roots of the early conservationists who situated humans above nature and strove to conserve natural resources for their later economic use. In contrast are the preservationists and later environmentalists who situate humans within nature as a part of the broad web of life and creation. In this view nature and wildness are to be preserved, protected from the effects of human activity, not for later development and economic use, but because they are sacred. In secular terms preservationists or environmentalists would argue an ethical basis for preserving nature -- destruction of "nature" ultimately leads to the destruction of humanity itself.

Whether one believes that humans as a species are above or a part of nature does not necessarily affect one's decisions on how to behave towards nature. It may affect <u>why</u> one makes a certain decision, but the outcome may be the same as we broaden our perspective about the extent, duration and kinds of influence humans have on the environment and the environment's influence on human well being. Especially when one takes into account the effect of current activities on the health and well being of future generations, (consider high level radioactive waste disposal problem, eroded farmland and silted rivers, and the staggering loss of primeval forest in North and South America) it is only the most technological optimist who would argue that the future will necessarily be better than the past.

Evaluating Environmental Benefits and Costs

The second question was how do we (meaning how does a politico-economic system) evaluate the costs and benefits of environmental protection or destruction. The most narrow interpretation of this question can be illustrated by two antagonistic positions -- proffered of late in newspaper editorial pages. We read on the one hand that environmental protection is already too costly and that further requirements such as those imposed by the new Clean Air Act amendments are an unnecessary drag on U.S. competitiveness and the economy. On the other hand we hear that investing in pollution prevention and clean up is an appropriate displacement of investment from polluting (harmful) activities to pollution prevention activities and that this displacement results in positive economic growth and enhances our competitiveness given the broad international interest in (and ultimate investment in) pollution control and prevention. Which position is correct? You would be surprised I'm sure, if <u>I</u> did not subscribe to the latter characterization. But what is important is that <u>you</u> are able to think through the problem posed by these two statements and decide which, if either, of the positions you agree with.

What is behind these two broad conclusions? They both incorporate a number of important, albeit implicit assumptions. To simplify we can assert that the first statement assumes that little or no additional value or benefit would be gained by further environmental protection efforts. The second position assumes the contrary -- that expected benefits, however measured, exceed the expected costs. Implicit in these assumptions are questions about how we measure or project expected costs and benefits of environmental and health protection. Should we take into account value to future generations, and if so how do we do it? Does a particular environmental policy ask today's generations to forego economic well being for the sake of future generations or does the policy simply require more careful husbandry of resources to the benefit of all generations? Given that the expected benefits include public health protection -- reduced illness and mortality rates due to, for example, exposure to air pollution -- how do we value human life when comparing costs and benefits? And, apropos of my first broad question, how do we, or do we, value the intrinsic nature of human life and other things such as wilderness? Are we limited to tallying up a person's expected dollar contribution to society and the identifiable use values of virgin forests and spotted owls or can humans and or nature claim protection just because they are there?

One of the reasons it is so difficult to reach consensus on the answers to these questions is that we lack good tools with which to explore them. For example, conventional measures of economic well being such as the Gross National Product (GNP) measure the level of "market" activity. Much of our economic, material and spiritual well being is supported by non-market goods and services and is thus not reflected in indices such as GNP. Thus, if we base our health and environmental policies on their predicted impacts on GNP, we may under- or over-estimate the actual impact on society. The work of Herman Daly and John J. Cobb in For the Common Good: Redirecting the Economy Toward Community, the Environment and a Sustainable Future for example, provokes a reconsideration of what exactly our economy is and challenges our conventional measures of economic well being. Conventional indices do not account for the products and services provided by natural systems, nor for the economic losses sustained when pollution and land use practices damage those systems. They also do not, for example, distinguish between expenditures on heart surgery on the one hand and pollution control investments which reduce people's exposures to carbon monoxide, a contributor to heart disease, on the other. Daly and Cobb have constructed a new index that explicitly accounts for the effects of pollution on health and the environment, and the services, including future services, provided by natural systems.

Within the government bureaucracy, measuring and comparing the costs and benefits of various environmental protection proposals and programs is common place, in fact it is required under Executive Order 12291 for all major rules that a federal agency proposes. This "regulatory impact analysis" exercise is often performed perfunctorily however, with only the most obvious and easy to measure attributes carried in the calculations. Too often for example, categories of benefits that are difficult to measure, quantify or monetize are carried in the accounting as zero. However, even without broadening or revising the way we account for "goods and bads" in our politico-economic system the way Daly and Cobb and others are suggesting, an assiduous accounting of expected benefits of environmental protection requirements reveals dramatic environmental and health costs under today's laws and practices.

To illustrate, I turn to the Clean Air Act amendments. What are the benefits to be gained under the new Clean Air Act and how do they compare in magnitude to expected costs?

Benefit/cost analyses can help society, or the bureaucracies charged with carrying out society's wishes, in evaluating whether, and to what extent, various environmental protection measures should be undertaken. For example, President Bush drew a line in the sand early in the Clean Air Act amendments debate. Last January, roughly six months after he submitted the Administration's own Clean Air Act proposal for Congressional consideration, he informed the Congress in no uncertain terms that he would veto their CAA amendments if the costs exceeded by more than 10% the costs estimated for the Administration's bill. Implicitly the message was that the Administration had balanced costs and benefits of air pollution control. What was the Administration taking into account in this balance?

Cost estimation for a program that for the most part will not take effect for 5, 10 and 15 years is a highly uncertain process. Bearing in mind the difficulty and complexity of predicting future costs, under consistent assumptions the House and Senate bills are forecast to cost approximately $25 billion annually in the year 2005, close enough to the Administration's original proposal. The estimates are based on the cost of doing the job using currently available technology. They do not take into account the potential for improvements in technology, process modifications and the substitution of inherently cleaner processes or practices to lower the cost considerably before the compliance deadlines arrive.

What benefit do we expect from this level of investment? As of this writing EPA is in the midst of a thorough review of the current state of knowledge of benefits from environmental regulations. In the paragraphs below I will describe the categories of benefits -- whether they are quantifiable and monetizable or not -- that we expect from the Clean Air Act amendments. I will illustrate on the one hand the real costs of not acting to further protect public health and the environment from the affects of air pollution and on the other hand, the type of challenges posed by our need to be able to account, at least qualitatively, for the full costs of living in a today's economy.

The Clean Air Act amendments include health protection from reduced exposure to sulfur oxides, hydrocarbons and ozone (smog). The benefits include reductions in both chronic and acute health effects as well as reduced mortality rates associated with exposure to particulate air pollutants. Estimating health benefits is a controversial exercise. The monetization of certain benefits such as reduced mortality or better well being (fewer days of sickness for exposed individuals) raises ethical issues. The Administration has used values that range from $2 million to $9 million per life in recent regulatory impact analyses. Once monetized, benefits that occur in the future are, according to economic practice, subject to discounting at various rates, the choice of which itself is very controversial. Discounting reflects the time value of money; we are most familiar with the discount rate as one component of the interest rate we earn on our savings or pay on our loans. However, it remains unclear that we should assume that a human life saved now is more valuable than one saved in the future. Do we accept a time preference of life?

The second category of benefits is improved visibility. The haze that perpetually blankets almost all of the eastern U.S. is caused by air pollution. Many of us have grown up with this haze and wrongly attribute it to our humid climate; but water vapor alone does not cause the scattering of light particles that we see as haze. We expect that the reductions in SO_2 emissions required under the acid rain control amendments will increase visibility by 20 to 30 percent in the east, an improvement that is both perceptible and valuable according to visibility benefits analyses. Economists have developed a number of analytical techniques to estimate how much individuals value clear vistas and to find dollar values for other non-market goods. While these methods have flaws and limits, the alternative of valuing non-market amenities at zero is clearly worse.

The third category of benefits is protection of the public from exposure to toxic air pollution. The amendments will require sources to prevent such emissions by installing collection devices or changing production practices. The air toxics amendments will reduce not only the expected cancer death rates from exposure to these pollutants, but also non-fatal cancers, reproductive and developmental effects, and diseases of the lung, liver, kidney, thyroid, nervous system, skin, blood and heart. The amendments will also prevent accidental releases of toxic air pollution which on average caused 265 fatalities and 900 injuries each year according to a survey of the period 1983-1987. The benefits of these requirements include not only prevention of injuries, illnesses and fatalities, but often increased efficiency in the production processes as companies maintain and operate their facilities better to ensure against accidental releases.

The amendments will require the reformulation of a broad range of consumer products like paints and solvents that currently contain volatile organic chemicals (VOC). These chemicals, either in use or when they leak, expose households and workers to extremely harmful levels of VOCS. Expected benefits in terms of improved health and worker productivity are enormous.

Finally, there are benefits to the environment itself. The acid rain control provisions especially are designed to improve environmental quality. These benefits are particularly difficult to measure given the complexity of natural ecosystems, but in qualitative terms we expect reductions in emissions that cause acid deposition to reduce current, and prevent future acidification of hundreds of lakes and thousands of kilometers of streams; we expect them to reduce the chances of long term changes in complex ecosystems due to changes in soils; we expect them to reduce the acidity of mountain clouds that are suspected of damaging trees and soils in mountainous regions; we expect them to reduce ozone concentrations in these regions as well. Damage to our cultural and historical resources such as civil war markers and the tombstones of our ancestors will also be reduced.

The last points I want to make on benefits respond to claims that the Clean Air Act amendments will damage the nation's economy, exacerbate the current recession, increase unemployment and damage our international competitiveness.

First, remember that our estimate of the cost of the CAA amendments, $25 billion, probably over-estimates actual expenditures, and we find no justification for industries' own estimates at two or three times that level. Second, the highest cost is forecast for 2005, the first impacts no earlier than 1992, hopefully well beyond the current recessionary trend. Even the peak year impact adds less than one-half of one percent to the expected GNP of $7 trillion by 2005. Third, we project job losses at less than one-tenth the amount that industry projects, and we forecast that these losses will be temporary and more than offset by increased employment in other sectors.

In fact, several economists forecast the creation of a significant number of new jobs in various sectors of the economy in direct response to the amendments requirements. For every $1 billion spent on air pollution control equipment, for example, up to 18,000 jobs could be created. The environmental industry presently accounts for roughly 2%, or $100 billion annually of the total U.S. economy (as conventionally measured) and employs roughly 2-3 million people. Thus expenditures on pollution control equipment or substitution of non-polluting inputs or products for traditional goods, is not lost to the economy but merely shifted towards less destructive activities.

As far as international competitiveness is concerned, industry actually ranks the costs of environmental compliance rather low in its list of factors considered when choosing where to locate new facilities, thus refuting the claims that raising the cost of doing business for American firms will drive them overseas. A comparison of net exports of goods across countries from major polluting industries shows little difference attributable to varying degrees environmental regulation -- primarily because environmentally related expenditures are small relative to total production costs. Second, the international market for pollution control equipment and inherently cleaner industrial facilities will only get bigger over time. US firms in this market must compete with Japan and West German firms, and the experience and competence they gain meeting the demands for cleaner air in the United States will help them compete to fill the demands in Eastern Europe and elsewhere.

The costs-to-benefit comparison exercise will play an increasingly larger role in the environment-economy debate. Thus it is very important that more people understand the terms of the debate, and the assumptions, ethics and values embedded in the various calculations.

Internalizing Environmental Costs

The use of market instruments such as emissions fees and tradable emissions permits is the most discussed environment-economy idea (at least in Washington). Marketable permits and emissions fees are devices useful in implementing environmental protection goals. Note that implementation follows society's decision that some level of pollution is too much and the state's decision to restrict pollution to some acceptable level -- the benefit cost comparison. That in turn follows society's expression of preferences for, and value of, an environment that is safer, ecologically less threatened and or more aesthetically pleasing. Only after we have articulated our environmental preferences and established our goals can we proceed with the decisions on how best to achieve the objectives of less pollution, cleaner and safer air, and ecological soundness. I dwell on this because so often I hear people talk about emissions trading and emissions taxes as if they are the environmental decision. In fact they are simply the preferred tools of the 1990's.

You will hear that just as centralized planning has been rejected in eastern Europe and the Soviet Union in favor of the market system because of the market's purported efficiency in revealing preferences and allocating resources, so too must the regulation of conventionally non-market commodities, such as environmental quality, be somehow brought into the market or price system. Where this is not possible, the politico-economic system must adopt new ways of valuing non-market items and reflecting such values in political and economic decision making. Both suggestions imply closing the gap between the price of an environmental good or service and the cost of that good or service. The acid rain control amendments are a good example of the first suggestion where the cost of SO_2 emissions -- acid rain, regional haze and other effects of pollution -- will now be reflected in the cost of generating electricity and therefore in the price of consuming electricity.

It is important to remember that non-market commodities, such as clean air, conventionally lay outside the price system for a reason and it is only by decision and action of the state that they can become incorporated into the price system. Such "common property resources" often have physical natures that make their common ownership inevitable, e.g. they may not be easily divisible or it may be difficult to exclude their use by some or all persons. Under our system of law many environmental resources are held in common (or in some cases unowned). Parks, wilderness areas, the air, roads, and many water systems are resources held in common. No nation to my knowledge has yet claimed ownership of the oceans although nations do restrict commercial activity in the oceans within some distance of their shores, such as the U.S. restricts fishing within a 200 mile limit. In many cases the use of common property resources is unrestricted, or under-restricted leading to their overuse and misuse.

Over the past 20 years, the U.S. government has increasingly restricted the use of air and water, limiting the location, amount and kind of allowable discharges under the Clean Air Act and Clean Water Act. Environmental statutes often limit discharges of pollution indirectly by creating a "command and control" system of environmental regulation. The government commands with a great deal of specificity and polluters control accordingly. This command and control system works. It forces firms to internalize, up to the level required by law, the environmental externalities they cause. For some problems, however, it is a very inefficient solution leading to higher control costs and lower levels of pollution reduction than might otherwise result. Thus we have the push for regulatory reform to incorporate, where ever possible, the price system in the regulation of environmental quality.

The Clean Air Act amendments contain many innovative economic incentives. The bill incorporates more economic principles than any other environmental law ever passed in this country, more I believe than are embodied in any country's environmental practices. The President and his immediate advisors are obviously strong proponents of the marketplace, and the original Administration bill attempted to bring the force of the marketplace to bear in our national effort to protect air quality. In doing so we gave momentum to an idea that has flourished in the academic literature for decades. In so doing we acknowledge (tacitly) (1) that the social costs of private enterprise are not factored into the cost calculations of private firms, and that the state must force private enterprise to internalize those social costs, and; (2) that the way to do that is, where appropriate, to use the force of the marketplace to rationalize firms decisions on how much to pollute and to ensure cost effective compliance with the bill's objectives.

In addition to improving the cost efficiency, the bill's use of market incentives, helps clarify the proper roles of the Federal government and the business community. It is the business of the Federal government, in response to the public's wishes, to set the nation's environmental goals, in this case the nation's clean air goals. The assimilative capacity of the atmosphere and oceans have been used as a free service to the firm, and as such, they have been freely destroyed. In the economist's terms, pollution is an externality of the production process and unless the state forces industry (and consumers) to bear the cost of pollution, industry will continue to exploit those resources until they are exhausted or ruined. Only the state can force economic actors to change their investment and production decisions to internalize these goods and services to reflect their value to society.

It is the business community's role to attain those goals in the least expensive, most cost effective manner. Indeed as a society we are collectively interested in minimizing the cost of producing all goods and services, including environmental quality. To do so we have to involve the creative, innovative energies of the affected industries. That is why the bill, particularly the acid rain control amendments, gives industry a great deal of flexibility in meeting clean air goals, and provides industry with economic incentives to take advantage of that flexibility.

Title V -- Acid Rain

President Bush declared during his election campaign that the time for study on acid rain had passed and the time for action had arrived. After years of analyzing acid rain control proposals introduced throughout the 1980's by various members of the House and Senate, EPA finally had the go ahead to design and analyze an Administration proposal. The purpose of the Clean Air Act acid rain control amendments is to achieve and maintain a 10 million ton reduction in SO_2 emissions relative to 1980 levels and to limit overall growth in NO_x emissions. A reduction of this magnitude will stop or slow the deterioration of our lakes, streams and forests that can be attributed to acidic deposition and allow most of those systems to start down the road to biological recovery. Moreover, it will improve visibility and reduce materials damage and health risks associated with these pollutants in the atmosphere.

The President's bill is the first major effort to incorporate market mechanisms into an environmental protection statute. With this bill we hope to demonstrate the contribution the market can make in reconciling the nation's environmental and economic objectives. The acid rain provisions combine a total and permanent cap or ceiling on allowable emissions while creating a market for emissions allowances under that cap. The pending legislation will motivate companies to find the best combination of control measures for their system. Once companies have determined how they will comply with the program, the market for allowances will allow those companies with high-cost reduction potentials to purchase allowances -- extra reductions -- from companies with lower cost reduction potential. The emissions allowance market will achieve a lower overall cost of the program because of the flexibility afforded to companies. Thus the amendments ensure environmental protection for years to come by requiring new sources to obtain emission allowances from existing ones. Newer sources have an incentive to be as clean as possible by not simply stopping their control effort at the level required under the existing law but through extending the controls if their cost of control is less than the market value of an allowance.

Under the acid rain control amendments, reductions in sulfur dioxide emissions would be achieved in two phases -- phase one reductions would be required after December 31, 1995 and phase two reductions would be required after December 31, 2000. The proposal also recognizes the benefits that this country and the environment can receive from improvements in our coal-burning technology and provides incentives, not only for demonstrating these new technologies, but also for deploying them as solutions to the acid rain problem. In recognition of their current state of development, this bill proposes to let sources which find certain repowering technologies to be a viable control option to receive a three-year extension of the December 31, 2000 deadline. In addition, those boilers amenable to cost-effective retrofit with low-NO_x burners would be required to install such devices by the end of year 2000. We believe that such a program, which is similar to other legislative proposals, will reduce NO_x emissions in 2000 by 2 million tons.

Consistent with our commitment to use market forces wherever possible, Title V calls for maximum flexibility in obtaining reductions. The plan would allow utilities to trade required reductions of sulfur oxides and nitrogen oxides so that they will be achieved in the least costly fashion. Roughly 9 million tons of SO_2 will be allowed to be emitted after the reductions requirements have been met. We created a new entity called an "allowance" - one for each allowable ton of SO_2. These allowances are distributed by statute among the existing utility units. Emission allowances can be traded freely within the entire lower 48 state region. Our analysis shows that the pattern of trades is such that large scale trading across regions is not really expected -- there will be some but most of the economic efficiencies are expected by trading within companies and between companies in close proximity. The trading system provides incentives for utility managers to choose the least cost solution to meeting environmental goals. It allows utilities to choose between fuel technologies and energy conservation strategies to meet their emission goals.

The combination of the emissions cap and trading provisions provides strong incentives for utility-sponsored energy conservation and stimulates innovation in both pollution control and electric generating technologies. It is also consistent with efforts to reduce carbon dioxide.

What kind of opposition and support did the acid rain control proposal attract? Predictably, we drew a barrage of industry opposition - let me call this brown opposition. Less predictably, we drew very little opposition from the environmental community green opposition. The main reason for this outcome is that fundamentally the acid rain control amendments are quite stringent, especially given the imposition of the cap. In many ways it was amusing to be arguing with executives of electric utility holding companies, those scions of industry, that the market could in fact work. Perestroika comes to American industry.

Industry characterized the cap and trading system as unworkable. They argued that it substituted a new cumbersome regulatory process, where each operating permit would have to be modified for each trade, for the old cumbersome command and control system. They argued that since (in their opinion) trading would not work, the cap would curb their ability to provide low cost, reliable electricity to customers. It took considerable effort on the part of the Administration to convince Congress that the emissions cap is technically and economically feasible, and should be an integral part of any acid rain legislation.

The trading system would operate like a checking account. The utility would be free to buy and sell emission allowances -- without amending the permit -- just as bank customers are required by the bank to hold sufficient funds to cover their checks. And just as a bank does not get involved in how to earn and spend the money in its customers' accounts, EPA would not tell a utility how to meet its allowance and emissions obligations. The bill's requirement that every unit install and operate a continuous emissions monitoring system would provide EPA with the necessary certainty about the location and magnitude of emissions.

By providing incentives for utilities to choose the least cost solution to meeting environmental goals and maximizing flexibility, the amendments result in the cheapest 10 million ton acid rain reduction proposal on Capitol Hill. A recent study by the electric utility industry agrees that a simple and flexible trading system would reduce costs by about $1 billion/year. In addition to reducing overall costs, trading would spread the costs more evenly across utility systems, thereby mitigating some of the rate impacts that might otherwise occur. Trading would also encourage innovation in pollution control technologies and reward utilities for pollution prevention and energy conservation investments.

The dearth of green opposition prevails because one of the leading national environmental groups, the Environmental Defense Fund, a member of the Clean Air Coalition, not only supports the emissions trading provisions, but actively promoted it within the Administration. In fact, I do not believe that the President's proposal would have included an emissions cap, even though it is intrinsic to a functioning allowance market, without EDF's efforts. What little green opposition we (EPA) encountered focussed mainly on whether we technically could run an accountable, enforceable allowance market, not on whether we should. I must add that we would have encountered more green opposition if we had not promised that the emissions allowance program would be in addition to, not in lieu of, other protection under the Act. I will return to this point later. The ethical or moral program of creating rights and a market for pollution was posed infrequently by the left, once in a New York Times op-ed piece by Todd Gitlin of University of California and again by Barbara Ehrenreich in Mother Jones magazine. Both pieces oppose giving industry the "right" to pollute, apparently on moral grounds; pollution is bad and no one should have the right to pollute at all lest we create other rights to do bad things. Both pieces use fables of a future full of rights to do bad things: street crime, white collar crime and international crimes against human rights such as torture. Their point seems to be that establishing pollution rights put us on a slippery slope towards regulating all "bads" through the market place and away from making absolute determinations of right and wrong by expressly prohibiting those wrongs.

I find their analogies inept. Emissions of air pollutants are not necessarily hazardous to human health or the environment. I know no responsible person who would argue that the only safe level of sulfur dioxide, for example, is zero. In fact as a general point, emissions are only pollutants when they exist in excessive concentration or in the wrong place such that they can cause harm. Just as a weed is a plant growing in the wrong place. Contrast that with street crime or torture. No one would argue that these are OK in limited quantities or in the appropriate place. More fundamentally, whether we say industry has emissions rights or not, we currently allow high levels of pollution to be freely emitted. Under the acid rain program, those allowable levels will no longer be free.

However, these critics are on point in their prediction that the tendency or inclination to use the free market to achieve any and every societal goal may subvert those very goals. My inelegant definition of a pollutant as concentrations of something in the wrong place or in excessive amounts is critical to understanding why I believe that it was appropriate to create an emission allowance market to regulate acid rain-causing pollution but why it is inappropriate to rely on it to regulate more local pollution.

The acid rain phenomena by its very nature has a large spatial scale. Acid rain causes or contributes to problems throughout eastern North America. Control strategies could be designed to focus deposition reductions on sensitive resources such as the Adirondacks or eastern Canada, but we know now that we need protection in the mid-Atlantic and southern states as well. As long as the magnitude of the reduction is sufficient, as our ten million ton goal is, we can be relatively indifferent as to the location of such reductions. Thus the efficiency gains expected from the allowance market will not make the program less environmentally protective than an alternative implementation method. Our assumption of spatial indifference is quite convenient from an implementation stand point. The task of regulating varying levels of deposition according to environmental sensitivity, and allowing emissions trades that do not violate those regulated levels is daunting.

Certain other environmental problems may lend themselves to an emissions trading system. Global warming caused by a build-up of greenhouse gases, seems to have the right attributes: a spatially large scale phenomena, caused by emissions of pollutants that can be measured and accounted for, emissions that are controllable through actions that can be described, agreed to and enforced. All it will take is a global commitment to limit the absolute magnitude of emissions of each greenhouse gas. Theoretically this is possible. Politically, it is difficult just as the acid rain problem was politically difficult for at least a decade.

I do not believe that the same can be said for many other environmental problems, especially air pollution problems, and I would discourage people from believing in a brave new world of the market-based pollution control strategies. In fact I would caution them to look at all such proposals very skeptically. Unless a trading scheme accompanies a large and measurable commitment to reduce the pollutants of concern, and unless the scale of the problem is large enough to support a competitive market in emissions rights, the trading scheme may be a shell game.

For example, we have made tremendous strides in the past 20 years protecting public health by setting and enforcing uniform national standards that prohibit excess concentrations of air pollution and uniform national standards of allowable emissions at new facilities. The national ambient air quality standards for the two pollutants that cause acid rain, nitrogen oxides and sulfur dioxide, are rarely exceeded within the United States. The 1970 and 1977 amendments to the CAA provide good protection of public health from the local impacts of these two pollutants. But the earlier law did not provide protection from the problem of transported air pollution, thus the acid rain control amendments this time around.

A national market in these two pollutants is appropriate in the context of the acid rain control problem but it will not protect public health everywhere if the limits on local concentrations are lifted. Excess local concentrations are often caused by emissions from just one or a few sources and the ambient standards must be strictly enforced to continue our good record of protection. Local protection often means case by case reviewing and setting of allowable emissions levels, a situation mistakenly considered antithetical to efficiency in the minds of some who will insist that such case by case review interferes with the emissions allowance market. But recall, the allowance market is only legitimate in the acid rain case, it is inappropriate to local SO_2 pollution problems. Not only that, emissions trading would not necessarily lead to greater efficiency because efficiency only follows when a market is workably competitive. Local air sheds are often dominated by one or a few sources, not enough to make the market competitive.

Summary

Each public decision on how to use natural resources -- including mineral exploration and use, management and or protection of natural ecosystems and the assimilative capacity of the air and water -- implicitly or explicitly assumes the answer to three questions. 1) Are these resources valuable; 2) how valuable are they; and, 3) what economic or regulatory system best ensures proper reflection of their value in decisions on their use? These are public decisions because many natural resources are non-market goods. They are non-market goods because either they cannot be, or by state decision, are not, goods that conform to the price system. Considerable effort must be extended to focus our thinking on the first question, to provide the analytical tools to answer the second and to develop a set of regulatory tools in response to the third.

The Energy Outlook

Thomas P. Sheahen

Dr. Thomas P. Sheahen is a member of the Technology Evaluation Group of the Energy Systems Division at Argonne National Laboratory, currently emphasizing the study of high-temperature superconductors. Dr. Sheahen earned both his bachelor (1962) and Ph.D. degree (1966) from the Massachusetts Institute of Technology. Sheahen's career spans both technical research and government policy formulation. From 1966 to 1973 he was with AT&T Bell Laboratories, developing infrared measurements for energy conservation in industrial factories. As a Congressional Science Fellow (1977-78) and a Senior Policy Analyst with the Office of Technology Assessment, he worked on legislation related to America's energy conservation agenda. He served as Executive Director of the Energy Research Advisory Board of the Department of Energy (1985-86), and heads his own science consulting firm, Western Technology Inc. He is a registered Professional Engineer in Maryland and an adjunct faculty member of the University of Maryland.

Abstract

Perhaps the most important way in which humans interact with the environment is through energy -- both its extraction and its use. Because of their environmental impact, hydrocarbon fuels are particularly worrisome. This paper strives to provide an overview of our contemporary energy situation, enabling conference discussion to begin from a common starting point. Both the global and United States energy use patterns are reviewed, with emphasis on the past 20 years. The difficulties associated with making predictions about energy are explained, and no detailed predictions are offered. Some general trends, which may plausibly be expected to continue into the future, are recognized.

I. Framework

At this ITEST conference, our purpose is to discuss questions pertaining to the environment. Because mankind's use of energy has a major impact on the environment, and because energy-use correlates strongly with quality of life, it is inevitable that some of our questions deal with the topic of energy. Accordingly, this text begins by presenting a series of energy related questions.

Our most fundamental question is, how will energy matter in the future? There are two subordinate questions here: How will energy affect human life and the environment? (both globally, and in America); and Will there be enough energy? This latter question in turn raises two more: How much is enough? and Where will we get it from?

The pathway to the answers to these questions contains many more questions: What would the ideal situation be? (Energy is clean and cheap; everybody has enough; the underdeveloped countries get better.) For various socio-political scenarios, what is the likely outcome going to be? Will energy shortages limit the quality of life? Will energy supplies be sufficient to sustain normal growth? Will energy effluents degrade the world environment?

Certainly one class of important questions deals with mankind's stewardship of energy: We already know that high quality of life correlates with high energy usage; but can wise energy management strategies accelerate improvements in global quality of life? All such questions are special cases of the more general question: What can we do to make it better? As Scientists? As Americans? As Christians? If it weren't for questions of this type, there would be no particular reason for ITEST members to come together to discuss the environment.

The following pages address six specific topics: section 2 describes what is known about energy today; section 3 tells how our knowledge is used, and how it is limited; section 4 covers the art of predicting the future, with examples of what happens when you get it wrong; section 5 presents general guiding principles about energy in the future; section 6 observes some trends in progress; section 7 concludes by offering a few specific propositions.

II. Current Energy Information

A. Units: First of all, we have to agree on what we are talking about, and to this end we select a UNIT of energy. The QUAD is a particularly useful unit for global discussions, because the world uses about one Quad per day. The United States uses less than 100 Quads per year. The definition is that a Quad equals one quadrillion British Thermal Units (BTU). One BTU is the energy required to heat one pound of water one degree Fahrenheit, and one KiloWatt-hour (KWh) equals 3413 BTU. Note that KiloWatt-hours are a unit of ENERGY, whereas the Kilowatt itself is a unit of POWER, or energy per unit time. One Joule per second equals one Watt. A typical nuclear reactor or major coal-burning power plant produces about a Gigawatt (GW), that is, a billion watts = a million kilowatts.

Certain fuels deliver a fairly constant ratio of energy per unit mass. Natural gas, sold by the cubic foot, and oil, sold by the barrel, can be expressed in "energy" units. "Barrels-of-oil-equivalent" is then an ENERGY unit, and "Barrels-per-day" is a POWER unit. Outside the United States, "million tonnes oil equivalent" is a more popular unit than the Quad; 1 Mtoe = .04 Quads. When burned in an electric power plant, however, a million tonnes of oil produces only about 4 billion KWh of electricity, due to combustion inefficiency. All sorts of other conversion factors are routinely used by engineers active in the energy field.

B. Data: Some aspects of our energy knowledge are in very good shape. The oil companies have excellent data for both production and demand; Figure 1 is a typical example, from the BP (British Petroleum) Statistical Review of World Energy. On the exploration side, oil companies retain great quantities of raw geological data. For example, at ARCO in Dallas they have a warehouse full of magnetic tapes containing data on drill-cores, and a Cray to analyze the data with; all that is missing is software to correctly interpret all this data.

World production of primary energy over the past 20 years is shown in figure 2, which gives the breakdown into types of energy sources. Clearly, fossil fuels are about 90% of the world's energy supply. For the United States alone, the pie chart of figure 3 shows the breakdown for the year 1988; fossil fuels are 87% of our energy supply, down from 90% in 1978. The "energy balance" of the United States, which clearly indicates the gap between production and consumption, appears in figure 4; the height of the "imports" bar reveals our dependence upon foreign oil.

On the demand side, the U.S. Energy Information Administration collected extensive data between 1975 and 1985, and compiled a very good historical record since 1970 for the United States. There are at least a dozen good-quality academic studies around, which analyze one feature or another of that record. Where they disagree (usually about the future), it is because their input assumptions disagree.

C. Trends: One important fact that is widely agreed upon is that "Energy Intensity" is the way to measure progress toward properly managing energy resources. Raw energy data keeps going up, because the world population keeps going up. But the energy required per unit of production, or per unit of transportation, is a useful indicator of whether or not energy is being conserved. When automobiles get better average gas mileage, that's lower energy intensity, which shows progress. When people insulate their attics, it lowers the energy intensity of the nation's housing stock.

Taken together, the national index of energy intensity has been falling steadily since 1972, as shown in figure 5, even though total energy use has risen as irregularly as the Gross Domestic Product (GDP). This is caused by TWO effects: The most obvious one is conservation; but there has also been a shift in the mix of products vs. services in the economy. Microchips, the "information age", and so forth, have created a structural change, shifting the economy away from energy-intensive products. (For example, things made of steel are energy intensive, but computer

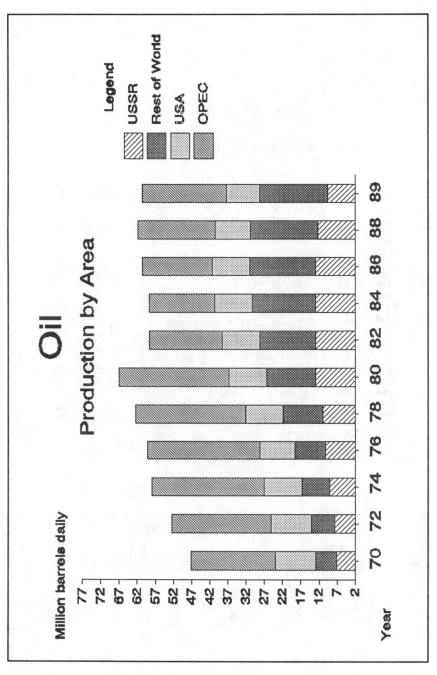

Figure 1 World Oil Production

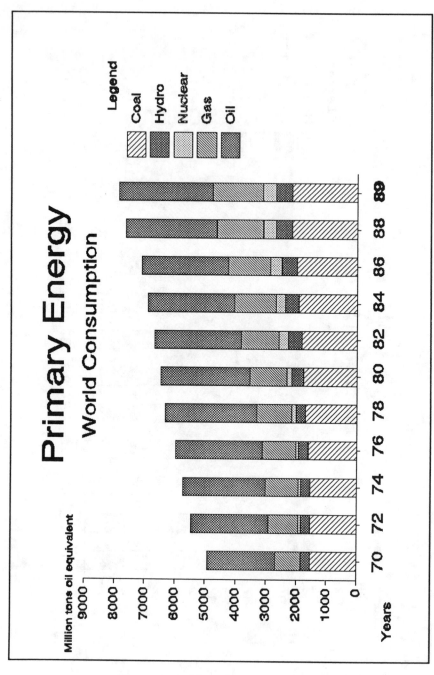

Figure 2 Primary Energy (World)

U.S.A. Energy Sources

1988 Production

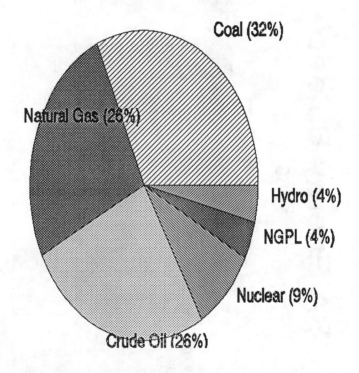

Figure 3   U.S.A. Energy Sources

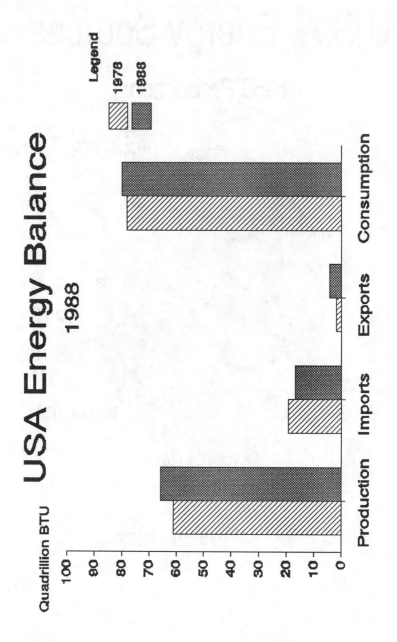

Figure 4   USA Energy "Balance"

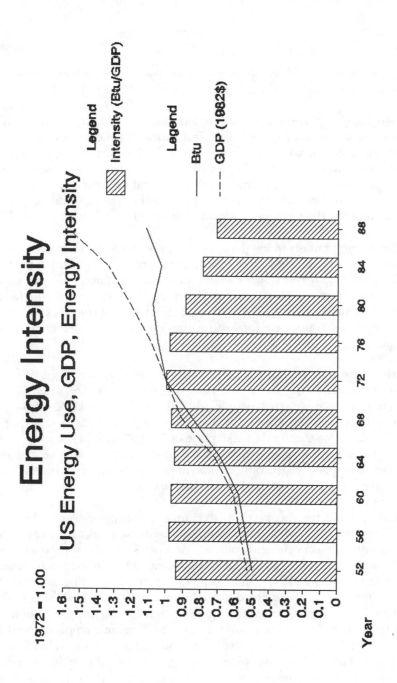

Figure 5 Energy Intensity

software is not.) This change is illustrated by figure 6, which shows the various causes of the reduction in energy per unit of GDP.

D. Missing Data: There are some weaknesses in our body of knowledge about energy. First of all, the historical record for the rest of the world prior to 1970 is weak, which implies that the baseline for "greenhouse" analysis is mushy. Second, the definition of "proven" reserves is a soft concept, leading to large error brackets around estimates of how much oil or gas can be extracted from a given field. Third, some countries prefer to keep their reserves information secret.

E. Economics: Energy obeys the laws of supply and demand, but there are always short-term lags. New inventions create new demands, but technology also drives down the cost of extracting energy; at the same time, dwindling resources run the cost up, but, conservation efforts mitigate demand growth. These statements have been true for centuries: energy commentators are fond of reciting the history of whale oil in the 1800s.

In the long run, supply and demand meet and set the price of each energy source. In the short run, however, demand is "inelastic", which means (for example) that people will fill their gas tank tomorrow no matter how high the price rises. Later on they'll think about buying a smaller car. We watched that take place in the 1970s and 1980s, as OPEC made huge profits suddenly, but later had a glut of oil. The call for government price regulation is motivated by the desire to avoid the consequences of short-term demand inelasticity.

Conservation is the cheapest way to counteract rising energy prices, and it is legitimate to speak of conservation as a "source" of energy. The Bonneville Power Administration, America's leading producer of hydropower, calculated in the early 1980s that adding insulation to homes was the best way to invest their capital, instead of building new power plants; they have implemented exactly such a program. In addition to the fuel saved through conservation, it is benign toward the environment. But the limits of conservation are also set by economic criteria: it makes sense to insulate your attic space, but nobody tears out their walls to wrap insulation around hot water pipes. Throughout America, utilities created various innovative pricing methods to encourage their customers to conserve. Factories usually responded faster than homeowners, and have engaged in "cogeneration" to reduce energy demand.

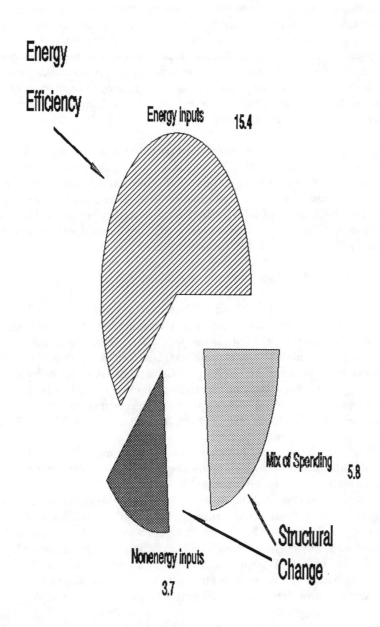

Figure 6 Reduction in Energy Use (Quads), 1972-85

Capital construction costs limit the energy supply in most of the world. There are many sites where hydropower or solar power are technically feasible, but the money to build the structure just isn't there. This is especially true in the underdeveloped countries. Any investment in energy supply must deliver enough revenue to pay off the capital cost as well as the day-to-day operating cost. When America built its dams in the 1940s, 50-year bonds at 2% interest were floated. Brazil and India can't get those kind of terms today.

Finally, there are some energy sources that make good cocktail-party conversation, but which are so miniscule as to be negligible. These include wind, geothermal and ocean thermal sources. Over the last decade of American energy production (see figure 3), the total energy was around 60 Quads, of which the "other" category supplied 0.2 Quads -- "other" includes solar, wind, wood, waste, geothermal, etc. None of these have proven economically sound so far.

III. Data Quality And Limitations

A. Value: There certainly is sufficient economic incentive for energy producers to collect accurate data. The value associated with oil is not only economic, but political as well. The data since 1960 for the developed countries (OECD members) is shown in figure 7. (Figure 7 also shows that forecasts are not too reliable, but we'll come back to that point.) Somewhere back in the 1930s, America ceased to be the world's leading supplier of oil. Since the postwar recovery of Europe, the gap between OECD supply and demand has been filled by Middle-East oil.

What is important to notice is how OPEC was able to read charts containing similar data and adapt their strategy for maximum profit: In the 1960s, OPEC bided its time. In 1972, OECD supply peaked while demand continued to rise; and so in 1973 the OPEC cartel struck. The OPEC nations made a huge profit, and started a recession in the OECD countries. By the late 1970s, the North Sea and the Alaskan pipeline had come on line, and when the recession of the early 1980s came along, the gap between OECD supply and demand narrowed, leaving OPEC with surplus capacity, sometimes termed a "glut". President Reagan was entirely safe in deregulating oil prices at that time.

However, we are not trying to analyze world oil markets in detail here. My point is simply that some major political-economic decisions have

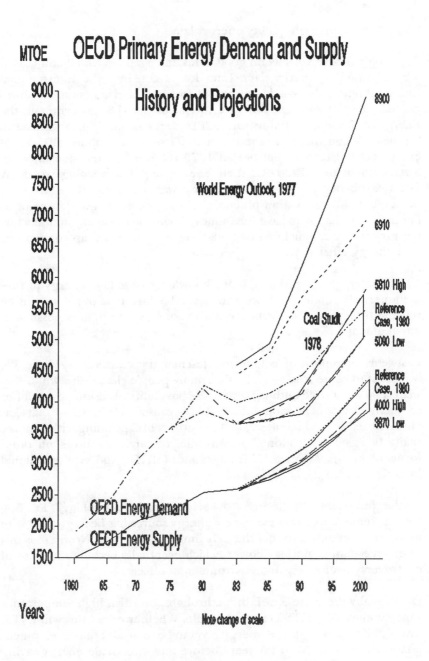

Figure 7 Energy Demand and Supply

been based upon high-quality energy data.

B. Sources: There are several good firms and agencies who do an honest job in compiling energy data. Data Resources, Inc., the International Energy Agency, the World Resources Institute, and others are well-regarded. Despite constant diffuse criticism of the U.S. government, the Energy Information Administration (EIA) is respected by professionals in the field. When the EIA required manufacturers to submit all sorts of energy-related data in the period 1979-1985, it had the side effect of forcing them to MEASURE their energy use. Upon doing so, many factories discovered how much money was going up the stack, and implemented conservation practices. They groused at the government, but they were happy to save the money. Today nearly every manufacturing process has a benchmark against which a company can compare its own energy efficiency.

C. Unknown Factors: Although our knowledge of today's energy picture is numerically quite solid, we can never be sure that our data will be reliable tomorrow. This is because a major breakthrough is possible at any moment.

Last year, the reports of "cold fusion" fascinated scientists everywhere. We wanted it to come true so badly that many people chased this will-o-the-wisp. The worldwide enthusiasm for the possibility of fusion power (after so many years of disappointing R&D) caused a blindness that has embarrassed several scientists, and left others backpedaling with "I never really believed it all along." Meanwhile, research continues on other forms of fusion, MagnetoHydroDynamics (MHD), and synthetic liquid fuels derived from coal.

In the June 1990 "Reflections on Science and Technology," Fr. Bob Brungs reminded us that every new energy source has been greeted with optimism, and only later did the ugly limitations appear. Nowhere is this more true than in nuclear power, which eventually became the focus of controversy over safety and environmental effects.

Occasionally there are REAL breakthroughs, of which high temperature superconductivity (HTSC) is an example. Whether or not this will EVER save significant amounts of energy is yet to be seen, and the development phase will occupy 10 to 20 years before any large-scale energy-saving applications will be achieved. Solar cells (photovoltaics) and coal gasification are now in long periods of development where economics limits their applications. Still, the door is always open to new technologies that save energy or convert energy efficiently.

Moreover, geology is a constantly evolving science, and better data analysis will improve reserves estimates. Advancing technology will allow poorer quality resources to be tapped, thus expanding the category of "proven reserves." One very puzzling but hopeful sign here is that extremely deep drilling routinely turns up natural gas in great quantities. The world's supply of natural gas may be limited only by the price of extracting it.

D. Changing Lifestyles: Not all breakthroughs are energy related. The transistor and its offspring (integrated circuits, microchips, desktop computers) set off a revolution that nobody predicted. One consequence is that driving to the office isn't as mandatory today as it was a generation ago; this allows energy savings. The major energy-effect, however, has been to shift the economy toward other activities that are less energy-intensive.

In the underdeveloped countries, the struggle for a better life includes a drive for more energy. Medicine is not energy intensive, but electricity is; and modern agriculture utilizes chemicals and machinery that require energy. Building a suitable infrastructure -- roads, schools, etc. -- demands great quantities of one-time energy input. We can sit up in America and deplore the way Brazil is trashing its rain forests, but remember that the people on the local level don't look at it that way. For them, it's "progress," because they haven't yet experienced the consequences of carelessness in a rush toward a more energy-intensive society.

Thus, while American lifestyles are shifting to slightly less energy use, most countries are rushing to MORE energy intensity. They're not inclined to back off before reaching our level of well-being. Our inability to control or even predict their behavior enroute to raising their standard of living is the most serious limitation today upon our knowledge of energy issues.

IV. Predicting The Future

A. Underlying Factors: Predictions of the future are subject to the input assumptions behind them. Once you understand the basis on which your own beliefs originate, you can begin to understand how other forecasters might disagree with you.

For example, if you say "Every American uses 1 KW continuously" (historically true), then your estimate of population growth instantly gives you an estimate of energy demand. If you assume an upper limit on total energy, then you can calculate what the upper limit on population must be. On the other hand, if someone else says that conservation can sharply lower the 1 KW figure, their forecast will allow the standard of living to rise without a concomitant rise in energy consumption, and they will get different limits for both total energy and total population.

The U.S. Congress' Office of Technology Assessment (OTA) has this to say:

> "Speculation about future energy use is fraught with difficulties and caveats. Factors that can be incorporated into a computer model tend to be insignificant in comparison to events that are nearly impossible to predict, such as the invention of the microchip or the Iranian revolution. Attempts at specific forecasts made in the mid-1970s accurately predicted that the energy intensity of the U.S. economy would decline, but underestimated the rate of the decline, leading to predictions that were 42% above actual use."
>
> -- from "Energy Use and the U.S. Economy", OTA Background Paper (June 1990).

Even the most eminent authorities are often deficient. Table I is taken from a National Academy of Sciences report of 1979, in which they forecast 1990 energy use in America. By imagining 5 different scenarios, each with and without conservation, the NAS was able to come up with bottom-line figures ranging from 61.4 to 139 Quads. Had you selected any one of their scenarios back in 1979, the spread would have narrowed to "only" 40%. (Incidentally, actual 1988 energy use in America was 80 Quads, as shown in figure 4.)

This section does NOT try to make better predictions than others; rather, it describes what goes into predictions, to explain why competent and

TABLE I: 1979 FORECAST OF ENERGY USE IN 1990

by National Academy of Sciences

	Total National Energy Consumed (Quads)	
	low conservation	high conservation
1974 experience	72.9	
Scenario A	88.2	61.4
Scenario B	105.3	72.6
Scenario B*	126.8	87.9
Scenario C	126.8	88.3
Scenario D	139.0	97.8

unbiased forecasters often differ. A typical example of forecasting discrepancies appears in figure 8, which compares seven different predictions for electricity demand for the 25-year period out to 2010, all made in 1985.

B. Oversimplification = Over-optimism: There is a long history of glib predictions and slogans about energy, all of which have proved wrong. The phrase "electricity too cheap to meter," now widely ridiculed, seems to have originated with the eminent mathematician John Von Neumann. As recently as 1965, Alvin Weinberg (Director of Oak Ridge National Laboratory and a leading pioneer of nuclear power) argued the case for water desalination, saying "The technological fix for water is based on the availability of extremely cheap energy from very large nuclear reactors." Weinberg made a number of additional predictions on the same basis of cheap nuclear power.

Bob Brungs dismantled this kind of thinking, citing examples where forecasters snapped up scientific advances but overlooked engineering realities ("Reflections on Science and Technology," June 1990). We still have this problem with predictions about controlled fusion, even though for nearly 40 years researchers have been saying (every year!) that we'll be there in 25 years. Engineers who work in coal-burning plants have a sobering slogan: "God invented tar to keep engineers humble."

C. <u>Supply</u> <u>Predictions</u>: Where fossil fuels are concerned, it is hard to tell what's under the ground. The distinction between "proven reserves" and "potential reserves" rests upon the shifting basis of geological science. Moreover, the amount of oil actually recovered from a well is perhaps only half of what's there; in order to do "secondary" recovery of additional oil, special techniques are required, and these are often expensive. Therefore, there is an additional "supply" still in the ground, which is inaccessible at today's prices.

Natural gas is a valuable fuel IF it can reach the consumer. But where there is no pipeline, it is worthless, and is not really counted in the supply. Throughout Africa and the Middle East, great quantities of natural gas are "flared," that is, burned off the top of stacks rising from wells, in order to get it out of the way of the oil. The very definition of the "supply" of natural gas associated with oil thus hinges on the capital investment needed to build a gas pipeline to the wells.

Coal, meanwhile, is easy to forecast: "Lots!". There seems to be about a 600 year supply of coal. More to the point is the matter of how to burn coal cleanly, and this is the subject of R&D efforts. Coal also can be converted into liquid or gaseous synthetic fuels (SynFuels), but the economics have been poor to date. Each time there is a significant rise in the price of oil, SynFuels find their way back into the headlines; at some price of oil, it will definitely be worthwhile to convert coal into liquid fuels.

The biggest uncertainty today in energy supplies has to do with society's willingness to ACCEPT any given energy source. Public relations count for much more than sound technology. Solar, wind and geothermal get good press, but nuclear power is the workhorse energy producer, and the only source that can replace a sizable fraction of fossil fuels. Nuclear substitutes for coal in providing electricity. It is probable that as greater attention is focused on carbon dioxide as a greenhouse gas, the public will come to accept nuclear power. The current research program to make new reactors that are inherently safe is progressing well, as described in my "Notes on Science & Technology" published by ITEST in June 1990.

Space does not allow an expansion of the case for nuclear reactors here. My point is simply that forecasters have to make ill-informed guesses about public acceptance of certain technologies in order to assign numbers to the energy supply. Little wonder that forecasts disagree so much.

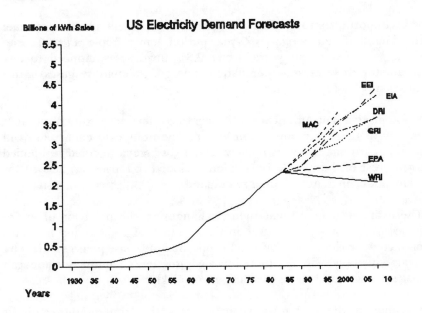

MAC -- Siegel and Sillen
EEI -- Edison Electric Institute
EIA -- Energy Information Administration
DRI -- Data Resources, Inc.
GRI -- Gas Research Institute
EPA -- Environmental Protection Agency
WRI -- World Resources Institute

Figure 8 US Electricity Demand Forecasts

D. <u>Demand</u> <u>Predictions</u>: The biggest uncertainty here is the growth of underdeveloped nations. Referring back to the American experience of each person using 1 KW continuously, you might predict that in N years 65% of the world will come up to that standard of living. Throw in your guess at world population and you've got an estimate of energy demand.

The "percentage growth" in population, in energy supply, etc., all are intertwined.

More important, small differences in ANNUAL percentage rates translate into huge differences over a 50-year period. For example, a growth rate of 2.2% triples in 50 years, while 2.8% quadruples. Applied to any parameter of interest -- population, energy, greenhouse gases -- this makes a colossal difference.

If you predict that Americans will learn to conserve energy at a rate that exactly matches our population growth, then our own energy use will remain constant. There is no way such a rosy scenario could be applied to the rest of the world. The underdeveloped countries are headed for tremendous increases in energy demand.

There are also occasional sudden changes in the behavior of entire nations. An energy forecaster in the summer of 1989 would not have predicted the changes in Eastern Europe, and the consequences still aren't clear: will everybody over there buy a gas guzzler? Changing societies can make energy predictions obsolete very quickly.

In America, the shift toward smaller cars, the "conservation ethic" is clearly a correlate of the environmental movement. No energy forecaster today knows how far that can be pushed. Our experience in the 1980s suggest that Americans have a very short attention span for energy conservation.

Still, Americans have definitely reduced their energy demand substantially compared to what had been expected years ago. Figure 9 shows the "NERC fan," a compendium of demand forecasts for electricity consumption in America. Every year from 1974 to 1983, the projections were lowered, but the actual use came out still lower. Recent experience suggests they finally "got it right": the actual 1989 number is 530 GW, right between the 1982 and 1983 prediction lines.

Long-range predictions never spot downstream recessions in economic growth; all a forecaster can do is choose a growth rate slightly less than what is called for at the moment, believing that inevitably recessions will decrease the long-term average. But those who must plan to have hardware functioning when it is needed -- electric utilities, for example, whose time horizon is about 10 years -- have to play it safe, and prudently overestimate demand. Referring to Table I, the very large spreads on each scenario make these numbers almost useless for utility planning purposes.

E. Consequences of Poor Predictions: When forecasts are wrong, what happens? Predictions have been wrong on either extreme. Two things we have seen in America are the overbuilding of nuclear power plants and the lines at the gas pumps. Such mistakes eventually are corrected, but it takes time. People switch to smaller cars, and electric generating capacity eventually gets absorbed. More gas wells are drilled after a winter with shortages. We have to accept the fact that forecasting is a very imperfect art, subject to frequent disruptions.

Meanwhile, the third world just keeps lurching along, with its energy availability subject to the vagaries of the developed world. These countries are a yo-yo on the end of a string driven by the "haves." One expensive error in predictions for America, Japan or Europe can create a shortage of capital reaching the less developed countries. Too generous a forecast quickly leads to credit problems, as we have seen in Mexico, who borrowed heavily when oil prices were high.

For those of us in America who feel a special kinship with the Missionaries in far-off lands, it is particularly disheartening to realize how much the future of those countries depends on the global politics of energy. They get the crumbs from the table, especially in terms of the capital needed to deliver energy to their own people. (Mountainous islands like Haiti and Borneo have adequate geographical sites for hydropower, but no electrical transmission network.)

With hindsight, we can observe that construction of the Alaskan pipeline during the 1970s, which was opposed on environmental grounds at the time, helped to drive OPEC into glut conditions, and lowered the price of oil to the poor countries, thus giving them a better chance to develop. Absolutely nobody foresaw that outcome in 1974, when the pipeline decision was debated. This promotes the temptation to try to justify further exploitation of our own resources, but such reasoning is specious: OPEC has gotten smarter in 20 years, and short-term fixes only leave us with a worse problem when our resources start to dwindle.

Historical and Current Electricity Demand Projections

The "NERC Fan"

Summer Peak Demand Projections
Comparison of Annual Ten-Year Forecasts
NERC -- U.S.

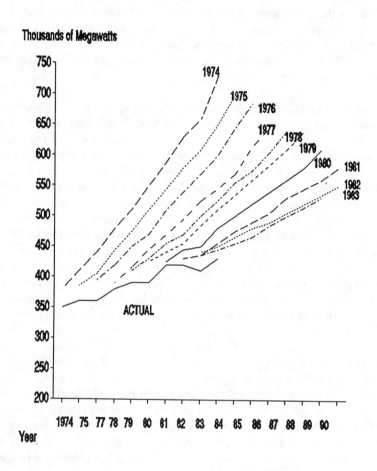

Figure 9 Electricity Demand Projections

V. General Guiding Principles

A. <u>American</u> <u>Scenarios</u>: Each American uses about 1 KW of electricity continuously. (Built into this is the electricity used in making steel, in the subway systems, etc.) Anyway, the electric demand is directly proportional to the population. If one further assumes driving patterns remain the same, then the average adult will continue to drive about 18,000 miles per year, getting 20 miles per gallon. This works out to America consuming about 2.5 billion barrels of oil this year, or around 14 Quads in gasoline alone. If you tack onto these figures a 2% annual growth rate (in population, in the economy, etc.), you can generate figures 22% higher for the year 2000, or 48% higher for 2010. The numerical details are not really of interest; the message is clear that "somethin's gotta give." The least painful path for America is to take conservation VERY seriously We have done well since 1973, but conservation still has plenty of room to grow.

B. <u>Rest</u> <u>of</u> <u>the</u> <u>World</u>: Many countries wish they had America's standard of living. World energy consumption has already risen 60% in the past 20 years, as shown in figure 2. The worldwide recession of the early 80s put a sharp dent in oil consumption; only in 1989 did oil recover to its level of 1978 & 79. Clearly the trend is steadily upward, although a 60% gain in 20 years is only a 2.4% annual increase. Continuing that rate for 100 years gives a multiple of 10 increase in energy demand.

How fast the other countries will advance is anybody's guess, and your particular guess can be used to construct world energy estimates for the year 2000, 2010, etc. However, the only prudent way to do long-range energy planning is to suppose that they will eventually reach America's level. If you select some distant date (year 2100?) for the world to reach our level asymptotically, then you have to explain how you will supply 1 KW to 10 billion people. One facile calculation yields the answer "with 10,000 nuclear reactors." That implies building 100 nukes per year every year in the next century; so far America has built just about 100 nukes total in 25 years.

Again, somethin's gotta give. In all likelihood, the poor nations will remain in squalor for many centuries. This is neither the humanitarian, the Christian, nor even the American way to do business; it is merely the most likely outcome. A second possibility is that the world will embrace a new generation of benign nuclear reactors; that would require a

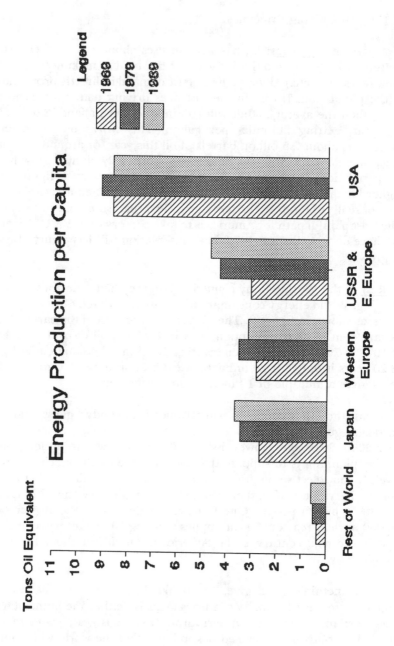

Figure 10 Energy Production per Capita

dramatic shift in attitudes, but only modest advances in engineering technology. A third pathway, certainly much desired, is that we will invent controlled fusion and it will turn out to be the panacea.

C. How NOT to do Planning: Here is a very crucial point which we, as scientists, cannot let political leaders forget: you must never do planning based on what isn't there. History is littered with the dead bodies of those who forgot this point. It is simply not credible to say to the third world in 1990: "Trust us. Back off now on energy and resource development and someday we'll come up with the perfect answer."

Don't count on controlled fusion! After 40 years of research, we are still "25 years away." Fusion salesmen plot progress on 10-cycle log-log paper, to disguise the fact that they're still a factor of 50 below break-even. The nearest working fusion system is 93 million miles away, and its containment system is gravity, not magnetism or lasers. As devoutly as we WISH that fusion would come true, and quickly, it is silly to count it in our energy plans.

Wind and geothermal energy likewise will not add up to much. Hydro is limited too, because there's only so much water running downhill. That leaves solar energy, which at present is not economic, but which may be in the future if certain engineering obstacles can be overcome. Large desert areas in the world offer unlimited opportunities for solar collectors, but it must be remembered that large structures are capital intensive. Any plan to rely on solar for the future must contain provisions for providing the capital.

D. Capital: Every important energy supply requires capital. Oil and gas sources must amortize the cost of a lengthy exploration period. Hydropower, which now provides 7% of the world's power, has been heavily financed with long-term government bonds carrying very low interest rates. Even conservation investments, such as better-insulated furnaces, require capital up-front. Contemporary investors will not wait long for their money to be returned, so favorable terms are scarce. Typical "market" interest rates, whereby energy investments must compete with all other choices of investment, discourage all long-term investments. This is a particularly severe barrier to energy development in the poor countries.

E. <u>Constraints</u> <u>on</u> <u>Planning</u>: Energy planning is constrained by parameters other than financial. The days in which major energy investments were done unilaterally are gone. The global economy links each energy source to every other one. Beyond the planning of capital requirements, the need to respect the ecology is by now widely recognized. There IS an environmental movement, and simply digging up or cutting down resources cannot be done without penalties. Every new major energy source must be approved by society. This is most pronounced in America, with public hearings, etc.; elsewhere pressure on governments by concerned citizens is starting to be felt as well. This means that society must be educated about energy, if only to avoid protracted fights with Luddites over unwarranted fears.

VI. Observation of Trends

A. <u>Conservation</u>: In America and the developed world, the cheapest new "source" of energy is conservation. Moreover, new methods of conservation keep turning up. Of course we can't reduce losses totally to zero, but there is still plenty of room to go. In the 1970s, certain conservation targets were established for American industry. These have all been exceeded, because new ways to save energy have been discovered that were unknown when the original targets were devised. New kinds of electric motors, new light fixtures, new types of automobile transmissions, all keep the "horizon" receding for energy savings.

Market forces drive products toward less energy intensity. Thinner beer cans and plastic shopping bags are familiar examples of consumer products modified by energy considerations. Every plastic component in your automobile saves energy in manufacturing and saves weight on the road. Many more everyday examples of such changes readily spring to mind.

B. <u>Structural Changes</u>: A quotation from the OTA summarizes this trend well:

> "Structural changes that result in less use of energy and the continued improvement in energy efficiency are likely to continue in the future. A driving force behind these two factors will be the continued development and diffusion of information technologies. Just as electricity generated tremendous energy efficiencies as it freed

factory design from the restrictions associated with steam and water power, information technologies hold out the promise for another revolution in the manner of production. These information technologies will place a premium on exploiting flexibility and the ability to monitor and control production to exact specifications, characteristics that are inherently energy-conserving."

Simply stated, technological changes in the industrial sector of the economy will tend to restrain the demand for energy.

C. Changing Lifestyles: There ARE feedback mechanisms at work that tend to conserve energy. Traffic congestion on highways and at airports promotes ideas which avoid travel entirely; the FAX and electronic mail are prime examples of this. Simple economics alone guarantees that such trends will continue as the price of fuels rise.

What we have NOT seen is any retreat from electricity use. Everyone gripes about power failures or even brown-outs, and blames the utility for not being prepared. Serious cutting down on MY personal electricity budget is not part of the ethic of many people yet.

D. Fuel Substitutes: Lack of capital has stagnated progress on oil substitutes. When considering a SynFuel plant, it is not a difficult calculation for an investor to discover that his cost is $50 or $60 per barrel, which won't sell on the open market. Since the United States government has curtailed subsidies for such projects, work has ceased on nearly all of them. Given the very tight current budget constraints, it would take a major change in public attitude (probably motivated by identifying "energy independence" with "national security", an idea that didn't catch on in the 1970s) to cause the government to resume funding energy substitutes.

E. Underdeveloped Countries: In these countries, energy use grows as fast as the capital supply allows. Hydroelectric dams, power plants, etc., are constrained primarily by limited capital. Often there are "institutional" barriers, such as no infrastructure (e.g., no electrical transmission grid), but these absences are traceable to capital shortages as well.

F. Nuclear Plants: There is a second generation of nuclear power plants now well into the development stage. (See the ITEST "Notes on Science and Technology" for June 1990.) Briefly, these reactors overcome all three of the major objections to nuclear plants: they are inherently safe, shutting down harmlessly when something goes amiss; they burn up their fuel efficiently, thus prolonging the uranium supply; and they recycle their own wastes, thus avoiding the need to store spent fuel for thousands of years. A demonstration reactor exists now, a full-scale unit is being designed, and such plants should be in routine operation by 2010. A second nuclear age, one that learned from the mistakes of the first, lies ahead.

VII. Conclusions

Hopefully, the reader now has an understanding of just how difficult it is to predict our energy future. Certainly one warning should be clear: Do not accept predictions based on simple linear extrapolations of present-day conditions. Perhaps what is needed now is to ask whether any light has been shed on the questions posed at the start of this paper.

Here are a series of propositions which I believe to be true. They make good "straw man" statements for the collected wisdom of ITEST to ponder. The information on the preceding pages offers a basis of energy information upon which a coherent debate can be structured.

1. Energy is the major variable in the "greenhouse" problem. No solution for the environment exists which fails to address energy considerations.

2. The next generation of nuclear power offers the best hope of overcoming the greenhouse effect.

3. The third world cannot long be kept in servitude; and they will demand lots of energy, which impacts the environment.

4. If America and the developed nations don't help the underdeveloped, they will try to solve their energy problems alone, and they'll do it all wrong: their solutions will be wasteful and polluting.

5. Education of the public is the chief task facing scientists. R&D is generally far ahead of the public level of acceptance. Education will be needed to achieve acceptance, and to achieve funding, from either government or private sources.

Finally (at the risk of undercutting everything I have said), it is worth-while to pause and inquire whether we are even considering the proper line of questioning. As an example of a challenge to my framework of thought, I offer this quote from syndicated economics columnist Warren Brookes:

> "At least part of the problem, it seems to us, is our very superficial definition for the word 'energy'. For most, it means some form of identified fuel -- oil, gas, coal, wood, uranium, solar light, wind, waterfall. But think beyond these obvious definitions to the real power that makes those fuels valuable.

> "It is safe to say that until the invention of the internal combustion engine, oil had little or no value and Arabia was a desperately poor land of nomads, herders and subsistence farmers. Until the discovery of nuclear fission, uranium was a largely irrelevant number on the chart of elements. Until the invention of the steam engine, coal was very nearly worthless.

> "Thus the real energy of our society arose not from raw materials in the ground but from the brain matter in our heads. It follows that the same brain matter that gave oil its huge value to our present economy should be able to take that value away from oil and give it to other more available resources."

Of course I don't agree with Brookes, or I wouldn't have written this paper. Nevertheless, the underlying question he raises has merit: Are we considering our energy (and environmental) problems on the right level of discourse?

Bibliography

BP Statistical Review of World Energy (BP America Corp., June 1990)

"Can Technology Replace Social Engineering?", A.M. Weinberg; in Technology and the Future, A.H. Teich, Ed. (St. Martin's press, 1986)

Energy Deskbook, Ed. by S. Glasstone, U.S. Dept. of Energy (Oak Ridge, 1982)

Energy Facts, 1988 edition, Energy Information Administration (June 1989)

Energy Use and the U.S. Economy, Office of Technology Assessment Background Paper (June 1990)

"The Future of Nuclear Power. Notes on Science & Technology," T.P. Sheahen, ITEST (June 1990)

"Reflections on Science & Technology," Fr. Robert Brungs, S.J., ITEST (June 1990)

"Reflections on U.S. Electricity Demands and Capacity Needs," P.D. Blair, presented at Aspen Institute Policy Issue Forum (July 18-22, 1990)

World Energy Outlook, OECD Publications Office (1982)

World Resources 1990-91, by the World Resources Institute (U.N. Environmental & Development Program, 1990)

Conceptions of the Human, Law, and the Environment

John M. Griesbach

John Griesbach received his B.A. in 1970 from Marquette University, Milwaukee, Wisconsin, going on to earn a J.D. in 1976 and an LL.M. in 1977 from the Harvard Law School. Griesbach, an Associate Professor of Law at the Saint Louis University School of Law, has taught in that institution since 1977. Among his courses are: Torts, Administrative Law, Legal Philosophy, and (occasionally) Environmental Law and Natural Resources Regulation.

The expression "the" environment is tricky. It suggests a single region that is everyone's environment. The definite article makes environment singular. Yet as a matter of history, there have been many environments, many accounts of nature and man.

John William Miller, <u>The</u> <u>Midworld</u> <u>of</u> <u>Symbols</u> <u>and</u> <u>Functioning</u> <u>Objects</u> (Norton, 1982), p. 84.

While I suspect that nearly all of us agree with Miller, his statement is a confusing claim. The trouble becomes apparent upon setting out some of the many senses of the word 'nature.' In one thoroughly modern sense, everything is part of nature, Chernobyl no less than the swallows of Capistrano. Grounded in our knowledge of causal continuity, this sense of the word admits no unnatural forces, no unnatural acts, no unnatural disasters. 'Nature' is a term of undifferentiated inclusion, with humans and what we do part of the causal structure of the universe along with everything else. To talk of "many environments" on this understanding of 'nature' simply makes no sense. There is only what there is, one environment, one nature.

There is a second, more colloquial sense of the word, however, according to which we distinguish the natural from the artificial. On this understanding 'nature' is a term of exclusion, used to denote everything uninfluenced by human hands. The criterion of use here is also causal, but it is a matter of causal discontinuity. It suggests that there are parts of the world that are what they are quite independently of the distinctly human. Insofar as we understand the environment to be natural in this sense of the word, it is of course true that "as a matter of history, there have been many environments"; it is simply to say that humans have influenced now more, now less. Yet this seems to miss Miller's point. His message is not that the environment has varied, but that what the environment is has changed as a consequence of "many accounts of nature and man."

This brings us to a third, far more venerable sense of the word 'nature.' We use it when we speak of the nature of the atom, the nature of life, the nature of man, the nature of government, and so on. Such talk expresses recognition that, while causality is continuous, it is a continuity of relatively closed systems. These systems exhibit a measure of what we might call causal integrity, a wholeness that evidences structure. Each of these systems, we say, manifests its own nature; each has its own properties and its own inner workings. But the closure is only partial, and relatively so. Every system is in some measure open, featuring as part of larger, more complex systems, both influenced by and influencing what is around it. These larger systems have their own properties and inner workings, their own natures. Yet they too are open. Some far more than others.

The most remarkable of systems are we humans. We are, at once, among the most closed and among the most open. It is our capacity to understand, to develop accounts of ourselves and of what is around us, and then to act on those accounts, regardless of other influences, that marks our closure. This cognitive inner-directedness, this freedom, distinguishes us, gives rise to our integrity, our wholeness; it manifests our human nature. Yet it is this same feature that has us among the most open of systems. We are able to reach out and to cognitively grasp, and so to be cognitively influenced by, all variety of systems -- from the intricately chemico-physical to the subtly interpersonal to the grandly cosmological. And as we are cognitively influenced, so also do we influence -- guiding and altering and fashioning. We are systems, yes, but system knowers and system builders even more.

It is ourselves in-context, context known and context made, I think, that Miller has in mind when he talks of environment. If so, as a matter of history, there have indeed been many environments, many anthropocentric complexes of systems. Each has been known and fashioned by our science, of course. But also by our literature and our theologies and our law.

My interest here is in environments that we know and fashion through our law. Like other modes of human discourse, law expresses conceptions of ourselves-in-context. When a legislature enacts a statute or when a judge issues an opinion, a portrait is drawn, perhaps of a response to a problem, or a song is sung, perhaps of a new way to get rich, or a story is told, perhaps of an oft-taken way of providing for descendants. What distinguishes legal discourse, however, is its extraordinary causal efficacy. By painting the portrait or singing the song or telling the story, the lawmaker initiates a scenario that tends to include just what is painted or sung or told. Thus law, more than any other human activity, can be seen as humans, directly and self-consciously, making themselves (and so also their contexts) according to their own lights. It reveals our conceptions of ourselves in context but it also comes to constitute part of the reality that accords with those conceptions. Moreover, it is a part that often endures long after the conceptions that give rise to it fade.

In this paper, I attempt to sketch three conceptions of the human-in-context that I think are both expressed and in some measure realized by existing law in the United States, and which shape much of the context within which we address what we see to be the environmental problems of today. One is an Enlightenment conception. It sees the human-in-context in terms of the self-defining individual entitled by nature to the land and other things with which he fashions his own personality. A second is utilitarian. It pictures the human-in-context in terms of elaborate systems of consumptive activities by which humans maximize their uses of resources under conditions of relative scarcity. The third is what we might call a structural conception. It sees the human-in-context in terms of conditions or structures, largely humanly fashioned, that in subtle and often unknown ways both limit and shape what we can be. As set out here, each of these conceptions is necessarily exaggerated and incomplete. What is more, since each of them, I think, captures part of what we humans are in context, it would be a mistake to suggest that any one of them has been held to the complete exclusion of the others. Nevertheless, because the focus of each vision is quite different, that part of our overall body of law that manifests each of them bears very differently upon our environmental problems of today.

Additionally, it is plain that there are other partial conceptions of the human-in-context manifest in law that bear upon today's environmental problems. Consider, for example, our readiness to conceptually organize ourselves, partly on linguistic and cultural similarity grounds and partly on the basis of geographic contiguity, which conception is legally manifest in our global network of nation-states. And further consider the difficulties that set-up poses for dealing with transnational environmental problems like acid deposition, global warming, ocean fisheries management, and ozone depletion. That and other conceptions manifest in law are relevant to the topic of this paper but beyond its scope.

I.   The Human as Self Defining Individual.

It is of course received wisdom that the basic structure of law and the lawmaking process in the United States is an expression of Enlightenment thought. Montesquieu's recognition of human diversity, and his pragmatic call to caution and compromise, given voice in the contentions of Madison during the 1780's, is plainly manifest in the numerous overlapping power-allocating provisions of the federal and state constitutions. We find the cynicism of Hobbes, with its complementary endorsement of strong central power, taken up by Hamilton and given effect in Article II of the U.S. Constitution, in its Commerce and Supremacy Clauses, in <u>Marbury v. Madison</u> and other early constitutional decisions. It is the legal realization of the conception most clearly set out by John Locke, of human as self-defining individual, naturally endowed with certain rights, however, that is of interest here. Locke's vision is, of course, enshrined in the Declaration of Independence and set out in some detail in the Bill of Rights. But it is as realized in the common law and in much of the legislation of the first 100 years of U. S. history that the Lockean conception shapes our approaches to today's environmental issues.

In this part of the paper, I briefly sketch the Lockean vision, with special reference to its labor theory of natural property, and then briefly describe its legal realization in the strong, undifferentiated individual property rights of the late 18th and early 19th centuries and in 150 years of governmental disposition of the public domain. I also set out a number of respects by which that Lockean-based law influences current environmental problems and our approaches to them.

A. The Lockean Vision

As is characteristic of Enlightenment thought, the Lockean conception takes the individual (more accurately, the white adult male family head) as the basic social, political and economic unit. This Lockean individual is basically decent, orderly, socially minded, and quite capable of ruling himself. He is an Aristotelian individual with a potential (or end) which he may not in fact realize, but which is in him to realize if his environment permits it. It is the individual's capacity to reason, for Locke, that enables him to fulfill himself. As he puts it in <u>Of Civil Government</u>, Ch. VI:

> The Freedom then of Man, and Liberty of acting according to his own Will, is grounded on his having Reason, which is able to instruct him in that law he is to govern himself by, and make him know how far he is left to the Freedom of his own Will.

Indeed, as "it is the understanding that sets man above the rest of sensible beings, and gives him all the advantage and dominion which he has over them," (<u>Essay Concerning Human Understanding</u>, Bk. I, Ch. 1, 1.) it is

"Reason" that gives rise to the individual's natural attributes or properties.

This analytic focus is perhaps most obvious in Locke's derivation of "the natural right to liberty." In another passage of Essay (Chapter XXI, 7.) he writes:

> Everyone, I think finds in himself a power to begin or forbear, continue or put an end to several actions in himself. From the consideration of the extent of this power of the mind over the actions of man, which everyone finds in himself, arise the ideas of liberty and necessity.

It is the individual endowed with "Reason" that commands Locke's attention. On introspection, it is the "power of the mind" as manifest in the individual's ability "to begin or forbear" that gives rise to "liberty" as a right natural to man. Notably, it is the positive liberty to do as the power of the mind directs, rather than the negative liberty to be loosed of the control of others, that the passage celebrates.

But it is Locke's theory of the grounding of private property that is of most interest to us. In Of Civil Government (Bk. II. ch. V., "Of Property,") he sets out the basic argument:

> As much land a man tills, plants, improves, cultivates, and can use the product of, so much is his property. He by his labor does, as it were, enclose it from the common. . . Nor is it so strange . . . that the property of labour should be able to overbalance the community of land, for it is labour indeed that puts the difference of value on everything; and let anyone consider what the difference is between an acre of land planted with tobacco or sugar, sown with wheat or barley, and an acre of the same land lying in common without husbandry upon it, and he will find that the improvement of labour makes the greater part of the value.

The conception is of individuals employing "Reason" through their labor -- tilling, planting, improving. Land, "without husbandry", and everything else without the application of labor, is valueless, or nearly so; it becomes valuable in virtue of the application of "Reason." Land and other things, insofar as they are valuable then, are the natural attributes or properties of the individuals who work them. What is more, it is by working land or by working iron or by working grain, that is, by extending his "dominion" over things, that the individual develops himself as a farmer or a smith or a miller. Again, it is noteworthy that it is the positive right, the individual's relationship to land forged through his active improving of it, rather than the negative right to exclude, that grounds the Lockean property right in nature.

Importantly, in the Lockean state of nature, uncultivated land and other unimproved resources are abundant. Thus, the allocation of property can be understood along the same lines as its origins:

> The measure of property Nature will set by the extent of men's labour and the conveniency of life. No man's labour could subdue or appropriate all, nor could his enjoyment consume more than a small part; so that it was impossible for any man this way to entrench upon the right of another or acquire to himself a property to the prejudice of his neighbor, who would still have room for as good and as large a possession (after the other had taken out his) as before it was appropriated.

(Of Civil Government, Bk. II., Ch. V, "Of Property") . Lockean individuals, manifesting "Reason" through their work, can each and everyone lay claim to the property whose value he creates. That property circumscribes a sphere -- entirely private and personal -- within which he carries out his individual self-development. Conflict of property rights, on this view, is logically impossible, a contradiction; either a man by his labor has made the land or other things he claims as his own, or he has not. Fortunately, there is enough, on the Lockean view, to take the labor of all.

Given this view of the individual and his natural rights, government is seen as a construct, fashioned by consent, and justified insofar as it safeguards those conditions necessary for individuals to realize their natural ends. Consent is in order as Lockean individuals are sufficiently reasonable to see that their well-being lies in mutual and peaceful cooperation. The powers of government, however, are narrowly confined. Primarily, it is charged with discerning and protecting individual spheres of liberty and property. Importantly, rights are natural to the individual, analytically prior to government, and so recognized or discovered rather than governmentally created or assigned. Moreover, extensive public ownership of land is simply not contemplated. Property rights are natural to individuals only. Government ownership is an artifact, a construct of consent, and so restricted to what is necessary to national defense, to the transport of mails, and to other activities appropriate to the common good.

Rather obviously, this Lockean conception of the human, particularly its natural right theory of property, addressed the everyday experience of 18th and early 19th century middle-class farmers, proprietors and tradesmen. In its country of origin, the English landed and entrepreneurs, who had exploited Locke's views shortly after the Glorious Revolution, were caught short by its implications as to the distribution of land and the product of industry. So they abandoned it in favor of Edmund Burke's glorification of stability, tradition and habit. But on this side of the Atlantic, where it well suited the experience of generations of settlers carving civilization out of wilderness, the Lockean conception took hold.

B. Strong, Undifferentiated Individual Property Rights

The extent to which the Lockean conception of a natural right to property influenced 18th and early 19th century judicial lawmaking in the English speaking world is perhaps put best in Blackstone's opening passage to his second book of <u>Commentaries</u> <u>on</u> <u>the</u> <u>Laws</u> <u>of</u> <u>England</u>. Sometime in the 1760's, he writes:

> There is nothing which so generally strikes the imagina-
> tion, and engages the affections of mankind, as the right
> to property; or that sole and despotic dominion over the
> external things of the world, in total exclusion of the
> right of any other individual in the universe.

The sentiment was not restricted to England. In the United States, the 19th Century treatises of Chancellor Kent, Joseph Story and .M. Cooley struck the same theme. Of course, it is by way of the legal details, most of which were fashioned as common law, that the vision was put to work.

One sees it in the diverging treatment of liability for personal injuries suffered as a result of collisions or medical treatment and liability for entry onto land. As to the collision and malpractice cases, the courts were in the process of developing what has become known as the negligence standard, a case-by-case assessment of the reasonableness of the manner in which the respective parties engaged in their various activities. As regards entry onto land, however, what we now call strict liability was imposed: every invasion, so long as the invader was aware he was on the land he was on (though not necessarily aware the land was the plaintiff's), was treated as a trespass, and redressable in suit for damages. The reasonableness of the defendant's entry was irrelevant, and even in the absence of actual injury to the land, the defendant was held liable for payment of nominal damages. A number of defenses were recognized, of which the most common by far was consent of the owner.

The Lockean theory of property is manifest. Land is taken as an essential part of the person who owns it. Any visible entry -- for any reason, regardless of damage -- is thus taken as an assault on personality. It is akin to battery -- the visible, direct attack on the human body. Each is an invasion of a constituent part (a natural property) of the self-developing individual. Consent, of course, was a defense for it indicated that entry was thought by the plaintiff to be compatible, and perhaps even part of, his own self-development.

Not coincidentally, the developing law of liability of landowners for injuries suffered by entrants took the same line. As to trespassers, landowners were liable only for killing or maiming, but even then only if the force used was in excess of that necessary to drive the trespasser off his land. As to social quests and others in whose presence the landholder acquiesced, liability was found for intentional injuries and upon failure to warn the entrant of hidden, artificial conditions of which the landowner was aware. Only in the case of injuries to business visitors was the landholder held to the standard of taking reasonable precautions to prevent injury. The Lockean notion of property rights circumscribing spheres of individual development with landholder responsibilities varying as entrants' participation in his activities vary, is apparent.

During the period, this Lockean notion of natural property rights was a central organizing concept, used to ground a great deal of lawmaking that in more recent decades has been treated as requiring a balancing of equities, or interests, or costs. Nuisance cases, for example, were treated as if the old Latin maxim *sic utere tuo ut alienum non laedas* (use your property so as not to injure that of another) was altogether descriptive. Litigating parties and their attorneys agreed that each landholder had a natural, prelegal property right to use, say, the airshed to dispose of fumes and odors. The disagreement was over the exact extension of the respective rights. The plaintiff alleged overreaching, that the defendant, by generating an unnatural measure of fumes and odors, had invaded his property (as with trespass). The defendant contended that the generation of fumes or odors was natural, and so part of his property right. The judge viewed his role as a matter of "finding" or "discovering" (as against "determining") the respective property rights, and, in doing so, looked to the "natural uses" of the respective parties. The opinion might take up the "reasonableness" of the parties' behavior, but the term was employed, not in the modern sense of balancing costs and benefits, but as involving an inquiry into the use of reason by the respective landholders in doing what they did.

So also with the law of riparian water rights that was developed under the leadership of Kent and Story during the 1820's and 1830's. All landowners whose land borders a natural watercourse were held to have certain natural rights to the water flowing therein -- rights of diversion, use, and consumption. The extent of each landowner's water rights were understood to be determined by the nature of the watercourse and by the nature of his enterprise. Diversion for a mill, for example, was part of landholder's property so long as the mill was situated on land adjacent to the stream and the water was returned to its natural channel before it left his land. The right to consume water varied with the needs of landholder's enterprises. Conflict was seen, again, as involving not a balancing of equities but the "discovery" of what water rights the parties actually had.

Interestingly, in consequence of this analytically prelegal notion of property rights, individuals in Lockean states of nature could jointly enjoy commons without the prospect of degradation from over-use. Each of many neighbors having a cow or two might use a common pasture, or the residents of a town might all fish a small lake for their own consumption. By grazing their cows and fishing the lake, the users create "value" and so acquire "natural" property rights in the pasture or the lake. These property rights, however, are limited by "Reason" to grazing by one or two cows per neighbor or to fishing only for consumption. Overreaching -- a neighbor's attempt to graze a herd of cattle on the commons or the initiation of a commercial fishing venture -- would thus be seen by all as violating everyone else's property rights. Thus, a sense of community, or moral suasion, or even self-help by the others could be expected to keep violators under control even without official recognition of the rights that were involved. Furthermore, in times of shortage or impending shortage, say, while drought was on or during spawning season, individuals might even be expected to recognize that their "natural" rights were reduced in measure as "Reason" required, and to "enforce" those reduced rights by custom and social pressure.

But this Lockean notion of "property" is manifest also in a great deal of judicial lawmaking not directly involved with land and other natural resources. Before 1850 or so, what we now call contract law was seen, not so much as a framework for the working of markets, but as a diverse grouping of decisions involving judicial oversight of the integrating and adjusting by individuals of their own Lockean properties. The vision comes through in the "meeting of minds" theory of contract formation, the use of Statutes of Frauds, in judicial readiness to distinguish among bargains on the basis of subject matter and to scrutinize the reasonableness of contractual terms, including price. In many ways, contract disputes were treated as analogous to nuisance and water rights and even trespass cases, with judges attempting to discover the natural property rights of the parties in detail and then looking into whether actual, reasoned consent was given to their reallocations.

Importantly, by way of the "due process" and "takings" clauses of the 5th Amendment and comparable provisions of state constitutions, this common law elaboration of Lockean property was given constitutional status, thus insulating individual activity from governmental as well as private interference. Individuals' natural rights to property, as "discovered" by state and federal judges in the course of trespass cases, nuisance cases, water rights cases, contract cases, and so on, were taken as the "property" which government was prohibited from "depriving" individuals of "without due process of law." "And "due process" in the "taking" of "property" required a condemnation proceeding, a finding of "public purpose," and the payment of "just compensation." The set-up essentially foreclosed federal, state, and even local governmental bodies from regulating or restricting land and other natural resource uses that were not also trespasses, or nuisances, or deprivations of water rights, etc., with respect to individuals. In this way, though not only in this way, was Locke's notion of limited government put into place.

The physical legacy of this Lockean conception of the human-in-context, influenced in no small part by its legal realization, can still be seen across

82

the landscape of the eastern two-thirds of the continent. Thousands of square miles of forest, and wetlands, and wildlife habitat have been "improved"; hundreds of rivers and streams have been locked, and dammed, and straightened; millions of tons of coal and iron ore and copper have been "given value" by the application of "labour." One ought not underestimate the extent to which the Lockean self-developing individual gave rise to our industrial and agricultural capacity, to our standard of living.

It is, of course, obvious now that the conceptions strengths were also in many ways its weaknesses. Land and other natural resources "lying in common without husbandry upon it" were "improved" by the application of the "Reason" of individuals who, as it happens, all too often paid too little regard to habitat and erosion and water and air quality. But the larger problem by far was the conception's failure to recognize, and so to fashion ways for dealing with, adverse effects of individual self-developing activity that were cumulative, indirect, non-obvious and long-term. The common law built on Lockean 'property' could not deal with resource degradation and loss that did not generate actual conflict in the "productive" activities of specific individuals. What is more, the constitutionalization of the Lockean vision disabled nonjudicial governmental institutions from stepping in. These disabilities were manifest not simply in an absence of power but, more importantly, in an absence of legal vision. Neither judges nor other public officials could contemplate that it was any business of government to limit or in any way regulate individual activities that either did not invade the property or other "natural" rights of individuals or that were incompatible with the narrowly understood "common good."

Though the Lockean theory of natural property rights no longer enjoys its central organizing position in legal thought and action, its spiritual legacy, in many ways, is still with us. Its notion of broad, undifferentiated, analytically prelegal individual property rights is part of our national psychology. It surfaces in the readiness of many to regard almost any governmental restriction or intervention regarding any use of any kind of natural resource as raising the prospect of every kind of limit or intervention as to every use of every kind of 'property.' Many of us, for example, would view a government agent's entry onto our private land to investigate hunting or waste disposal or agricultural practices as deeply analogous to his sneaking into our bedroom in the middle of the night. And consider also the difficulty in enacting legislation or promulgating regulations respecting natural resources that cannot easily be cast in terms of abating individually created nuisances. Whether the matter involved be access to beaches, erosion control, agricultural use of chemicals, rotation of crops, preservation of wetlands or wildlife habitat, public intervention is seen as restricting private property rights. In consequence, much of our law respecting natural resources has either been responsive and disaster driven or has entailed the use of monetary carrots that are always short in supply.

C.   Disposal of the Public Domain

While the U.S. judiciary busied itself elaborating the natural right to property during its first century, the executive and legislative branches were putting it into effect with respect to the millions who settled in the eastern two-thirds of the country. Recall that Locke's labor theory of value was meant to serve as a criterion for the distribution of land and other natural resources as well as the basis for the prelegal "natural" character of property rights:

> The measure of property Nature will set, by the extent
> of men's labour and the conveniency of life.

With vast territories acquired from foreign nations and Native American title reduced to the tenuous right of occupancy (itself an interesting bit of Lockean theory at work), the United States was in the unusual position of being able to put the distributional imperative into effect without inviting civil war and chaos. Indeed, the Property Clause, Article IV, § 3, ch. 2

> -- The Congress shall have Power to dispose of and make
> all needful Rules and Regulations respecting the Territo-
> ry or other Property belonging to the United States --

was understood, given Lockean notions of limited government, to authorize territorial governments and federal management of public lands only in anticipation of their disposition to individuals and the formation of new states. But again it is in the details, in the scores of statutory dispositions of the public domain, that we see the Lockean conception at work.

Most obvious was the recurrent statutory recognition that individuals fashioned their own property rights by actually working and improving land. Before 1820, at least two dozen special preemption statutes were enacted, authorizing settler-squatters to buy their claims at modest prices without competitive bidding, with liberal credit extended again and again. In 1830, a retroactive, one year general preemption act was passed. Another series of special preemption statutes were enacted during the following decade. In 1841, prospective preemption was authorized, enabling occupying settlers to purchase up to 160 acres for $1.25 per acre. In 1862, with the Homestead Act, payment for land was dropped. Settlers were authorized to claim 160 acres without cost upon settlement and actual cultivation for five years. In 1877, the Desert Lands Act recognized claims of up to 640 acres west of the 100th meridian at 25 cents an acre and proof that the land had been irrigated. In 1904, the Kinkaid Homestead Act authorized claims of up to 640 acres in Western Nebraska. In 1909, the Enlarged Homestead Act allowed claims of up to 320 acres (instead of 160) west of the 100th meridian without cost. And as late as 1916, the Stock-Raising Homestead Act authorized entry onto 640 acres of land "designated" as chiefly valuable for grazing.

Interestingly, where property rights were distributed by statute or by administrative action without regard to "the extent of men's labour and the conveniency of life," corruption and graft and ruinous speculation surfaced, and was deplored. Such problems plagued the early land auctions, the use of scrip to pay veterans, the land grants to new states and, notoriously, to the railroad corporations. And of course, each of the "labor"-based dispositional statutes was implemented with its share of fraud, and overreaching. Yet for all the land grabbing, it was not until the 1930's with the passage of the Taylor Grazing Act, that the Lockean distributional ideal was clearly replaced by a policy of government retention and management of the public domain.

Throughout this 150 years of western settlement, it was the strong, undifferentiated, Lockean property right that the settler obtained upon taking his land to patent. Not until passage of the Stock-Raising Homestead Act in 1916 did the federal government reserve subsurface coal and other mineral rights. Indeed, even those who took mining claims to patent under the General Mining Law of 1872, were accorded fee simple title, obtaining surface as well as mineral rights.

The legacy of this legal realization of the Lockean distributional ideal is apparent across the national landscape. It is obvious in the uneven distribution of public land as between the eastern two thirds of the country and the public land states of the West, where "the conveniency of life" (primarily the availability of water) raised substantial barriers to the fashioning of "natural" property rights. One sees it in the pattern of decentralized land ownership in the East and Midwest, with "family farms" increasing in size through the Great Plains and Texas, in the larger, more scattered private holdings of the West, in the worked out mines and ghost towns, and so on. As to the eastern two-thirds of the country, property allocation along the Lockean line exacerbates the problem of cumulative, long-term indirect resource degradation and loss. It accommodated widespread deforestation and loss of wildlife habitat in the 19th century, and the dust bowl of the 1930's. In some measure, it lies behind a wide variety of "tyranny of small decisions" problems of today, ranging from wetlands loss and ground water depletion to agricultural chemical runoff. In the public land States of the west, the view that the federal government holds the public domain in trust for disposition to Lockean individuals who can create "value" by the application of their "labor" remains alive. It surfaced in the Sagebrush Rebellion and privatization initiatives of the 1980s, in opposition to Wilderness Area designations, in resistance to putting offshore tracts off limits for mineral exploration, and in a host of other contemporary environmental controversies.

II.   The Human as Consumer of Scarce Resources

While the Lockean conception of individuals acquiring "natural" property by adding "value" to land and other natural resources builds on an assumption of abundance, the second vision of the human-in-context put into effect by our law is fashioned out of an appreciation of pervasive, though relative, scarcity. This second conception insists that all actual uses of things of the world -- time, human energy, land, other natural resources -- are always incompatible with other legitimate, but necessarily foregone uses. Every use of a resource, as the economist says, has its opportunity costs. On this view, individuals are seen, not so much to develop themselves in the course of their lives of work, but as consuming, using relatively scarce resources at the expense both of other users and of uses by others. Property rights are seen, not as "natural" attributes of the individual acquired through development of personality and "recognized" by legal actions, but as "positive entitlements" engineered and allocated by law in an effort to maximize total consumption.

In this part of the paper, I sketch in a bit more detail this second conception of the human-in-context and then set out two aspects of its legal effectuation that bear upon present environmental issues. It is of course obvious today that this utilitarian conception with its positivist view of law greatly undermined legal realization of the Lockean vision previously discussed. Blackstone's strong, undifferentiated individual property rights recognized and protected by the judiciary have given way in the 20th century to a much more complex set-up whereby diverse "bundles of sticks" are allocated, rearranged, and then reallocated by many different governmental actors. And the Lockean-based policy to dispose of the public domain has been repudiated in favor of federal retention and management for use. But my main focus here is upon two more general ways that law has effected the second conception. One has to do with its disposition to treat pollution and resource degradation as commons problems that can be solved only by either (1) assigning property interests to resource users who will then reallocate them to their "highest uses" by market transactions or (2) by direct governmental regulation of uses. The second general feature has to do with the readiness to employ "cost-benefit analysis" as the mechanism for governmental decision-making respecting land, natural resources, and environmental conditions.

A. Humanity Consuming

Utilitarian conceptions ordinarily distinguish humanity not into individual humans but into the numerous consumptive activities in which individuals engage. Some ground these activities in psychology, in drives to seek pleasure (including pleasures attendant to cognitive activity and altruistic behavior) and avoid pain; others with a more empirical bent speak only of wants and inscrutable preferences as giving rise to action. There is general agreement both that it is of the nature of humans to want, and that no particular wants or preferences are more "natural" than others. Wants are the sort of thing, however, that vary in intensity. Some want what is available from work more than they want leisure; others want current consumption more than they want future consumption. But given their ontological reductionism and this inability to distinguish in kind among an empirically evident diversity of wants, utilitarians generally contend that no cross-individual comparisons of wants -- no cross-individual "value judgments" -- can legitimately be made. Indeed, most contend that we cannot even compare cross-temporal consumptive activities engaged in by the same person: each action is grounded in the want or preference that was most intense at the time the action was taken.

It is the combination of great diversity in human wants varying in intensity together with relative scarcity that gives rise to utilitarian "value." Value is understood as exchange value: it is what actors give up or forgo in pursuing what they want. Work is viewed, for example, not as individuals developing their persons, but as an expenditure of scarce time and energy in exchange for commodities. Intrapersonally, it is disaggregated into the many commodities that scarce time and energy enable a person to provide for himself sleep, a pretty lawn, hours in front of the TV, a read book whose "values" are understood according to the personal preference curve revealed by his actions. Interpersonally, work is seen as an exchange of time and energy for wages or profits, in turn to be exchanged for other commodities; whose "values" again are understood as revealed by the various actors' actions. Rationality is assumed, but it is thought of, not as an exercise of "Reason" in identifying proper ends of life, but as acting, most of the time at any rate, in ways that maximize the satisfaction of those wants or preferences that a person happens to have at any particular time.

Obviously, in addition to disaggregating individuals into their want-driven consumptive activities, the conception disaggregates the things that people consume. Work, as just noted, is broken into expenditures of time and different sorts of human energy. Notably, land and other natural resources are also disaggregated into the many uses, often incompatible, that people want to make of them. A tract of forest, for example, is seen as many things -- a place for mass recreation and for solitude, as wildlife habitat and as a source of valuable minerals, a source of water supply and a place of fishing experiences. These many things are individuated according to the criterion of human "want"; and they are of more or less "value" according to differences in "opportunity costs" that are involved in using them.

Ironically, it is this disaggregation of things into their many, differently valued uses that has utilitarian thought, with its beginnings in the inscrutably private, driven to a vision of the active, regulatory state. As uses come into conflict with one another, they must be distinguished and allocated. Lacking a notion of "natural," prelegal rights, utilitarian thought has the state actively creating, extending and restricting all manner of "entitlements" (as against merely "recognizing" rights that are "natural" to individuals) . What is more, as "values" of various uses of things differ interpersonally and inter-temporarily, the state is charged with setting up and greasing the wheels of mechanisms that enable things to be put to their "most valuable" ("highest") uses, thereby maximizing the satisfaction of all wants. Thus, we get the two privileged coordinating institutions of the modern Western state: (1) governmentally structured markets and (2) governmental regulation under the direction of the democratic process.

Both markets and regulation according to the democratic process are justified as adding-up devices. With respect to the market, the aggregation is by way of supply and demand curves, thought to be fashioned out of people's varying willingness to pay for specific commodities. As to the democratic process, the aggregation is of votes into competing blocks of political interests, whose power is thought to be expressed in the readiness of elected officials to fashion regulatory programs that reflect constituent preferences. Market exchange, then, is seen, not in the Enlightenment way, as individuals rearranging their conditions of self-development, but as a wealth ("value") maximizing transaction. And legislating is seen, not as elaborating the good society by those inspired with republican civic virtue, but as a transmission belt geared to maximize satisfaction of the electorate's desires. Both mechanisms are seen as entirely impersonal, a valuable feature given utilitarian thought's skepticism regarding our ability to make substantive inter-personal value judgments. This feature, of course, is celebrated with respect to the market by use of the metaphor of the invisible hand; with respect to the legislative process, by various theories of collective choice.

Nearly all thoughtful proponents of the utilitarian conception acknowledge "imperfections" in working markets and in the actual operation of the democratic process. The invisible hand, they admit, is commonly arthritic -- with market outcomes greatly influenced by existing patterns of wealth distribution, distorted by characteristics of monopoly and externalities (costs and benefits accruing to non-participants), and frustrated by a whole series of defects which economists lump together as "transaction costs." Regulation pursuant to the democratic process is acknowledge to be just as "flawed"; distorted by defects in the electoral process, replete with "prisoners' dilemmas," and beset with implementation problems. Some extol the values of "liberty" over those of "equality" and, impressed with the flaws of the democratic process, put more stock in markets and in the prospect of adjusting to and correcting their

"imperfections." Indeed, the contemporary law-and-economics movement in legal theory, well-represented in the national government for the past decade, is committed to explaining
(and shaping) law in just these terms. others, valuing "equality" over "liberty" and more sensitive to market "imperfections", readily endorse direct regulation through the democratic process.

This utilitarian conception, with its mix of market and democratically coordinated allocations of resource uses is manifest in the bulk of environmental law put into effect in the 20th century. It is apparent in two general ways: First, in lawmakers' readiness to employ direct regulation only under circumstances where it is obvious that market coordination of resource uses is disabled by incorrigible imperfections; and second, in their readiness to employ cost-benefit analysis, a methodology designed to mimic the market, under circumstances where decision-making pursuant to the democratic process is thought to be beset with imperfections.

B. Regulating the Commons

A good bit of 20th century public land law is obviously grounded in appreciation of monopoly and wealth distributional flaws in market allocation. There's only one Grand Canyon. Water in the western United States is in critically short supply. Discovery and exploitation of oil, coal and other "non-precious" minerals, and lumbering, are skewed by existing technological economies of scale toward oligopoly and large accumulations of wealth. These market flaws have not invariably been treated as disabling. The most obvious instance is western prior appropriations water law, a blend of the Lockean and utilitarian conceptions, whereby water rights are allocated on a first-in-time/first-in-right basis upon actual "beneficial use," and where the rights are not subject to the riparian system's restrictions on transport or market exchange. But the overall legal approach, dating back to Teddy Roosevelt's creation of the National Petroleum Reserve and reservation of "pleasuring grounds," has been state retention of what's left of the public domain and regulation of use.

However, by far the most important utilitarian basis for direct regulation has come from its recognition that widespread and serious external costs are created as a result of market allocation of many uses of natural resources. This externality problem has contributed to the policy of retention and regulation of the public domain even more than has worry over monopoly and wealth distribution. Lumbering and mineral exploitation and grazing cattle on the public domain are seen to impose costs upon those who value clear streams and wildlife and solitude. Those costs are not internalized in the supply and demand curves for lumber and oil and beef because no one has been allocated property rights in environmental amenities--rights that if in existence would have to be purchased before the resources could be impaired. Additionally, those who enjoy environmental amenities are too numerous and scattered and beset with free-rider problems, etc., to get together to buy off resource exploitation. Market allocation of resource use, then, is "inefficient." Moreover, since those with wants not internalized in the market are in fact represented in the democratic process, it is not surprising that the mechanism of "collective choice" is employed in its stead.

In like manner, the recognition of externalities resulting from market allocations (and the high transactions costs in reducing them) has undergirded direct environmental regulation of more and more "private" activity. It underlies governmentally imposed fish and game limits, restricting even shotgunning of ducks on one's own land. It gets widespread and constitutionally pathbreaking deployment as a justification for urban zoning. It is used to justify extensive regulation of air and water polluting activities and of hazardous and toxic waste generation, transport and disposal. It underlies fisheries and coastal zone management, regulation respecting agricultural and developer destruction of wetlands and wildlife habitat, even bans on uses of DDT, leaded gas, and high-phosphate fertilizers.

Underlying all of these regulatory regimes is a recognition that some natural resource -- wildlife, the ambience of neighborhoods, the airshed, bodies of surface water, ground water -- is a commons. This utilitarian commons, however, is far different from the Lockean commons of joint, analytically prelegal ownership, with joint duties correlating to the joint natural rights. A commons to the utilitarian is a resource that is owned by nobody, something under a regime of universal privilege to use, takers-keepers, if you will, like the state of nature of Hobbes. Because no one has property rights in common resources, no one is able to demand payment prior to their use, no one can call upon the state to exclude. And with costs upon other users (externalities) not charged, the common resource is overused, degraded. This is no less the case with individual uses, e.g. driving ATV's in streams, use of plastic disposable diapers, than it is with market-driven uses, e.g. emission of SO_2 by power plants.

Furthermore, because both users who impose costs on other uses of common resources and users who incur those external costs (often the same people under different circumstances) are numerous, scattered, uninformed, unwilling to cooperate, etc., the buy-off or bribe is not available to prevent overuse. Thus, the democratic process has been enlisted to determine whether specific externalities of common resources use are certain enough, widespread enough, and serious enough to warrant direct regulation (with all of its weaknesses and with its restrictions on "liberty").

Yet it is becoming increasingly plain that there is a certain pathology to regulation of uses of common resources under the utilitarian conception. It is the view that all have privilege to use the commons though nobody has any right in it (and so nobody has any duties with respect to it) that gives rise to the need for regulation. Legal positivism, the position that rights and duties are created and assigned only by the state, undergirds this view. Regulation, as we have seen, is meant to prevent degrading uses by imposing duties. But there is a great deal of slippage between the wish and the act. Some of it occurs in the legislating process as "interest group politics" has representatives of every class of users scrambling for advantage. Much of it occurs because legislation and regulation promulgated pursuant to it is invariably vague and often ambiguous (a well-trained lawyer can find an ambiguity almost anywhere). Even if the language is clear, degrading resource uses can often be modified to take them out of the prohibitions. What is more, regulation must be enforced and penalties imposed for violation. And in these respects, it is not at all unusual to come upon the sentiments that unenforced law is not really law at all and that penalties are not meant to prohibit but serve merely to increase the cost of doing business. The background to all this, of course, is that same legal positivism that has commons as domains of universal privilege in the first place: with duties arising only upon state action, it is the measure of actual effective state action that determines the extent of the duty.

The moral legacy of the "natural law" thinking of the past is missing. With no notion of analytically prelegal rights, and responsibilities, common resources absent state action are treated as free-fire zones. The initiation of state action, however, only broadens the front of battle, with participants in "the democratic process" guided by no more than their constituent's preferences. Signs of the pathology are everywhere. It is apparent in legislative stalemate and complexity when all the deals are struck, in the incredible regulatory detail, in the litigation at every step, in the expenditures on investigation and enforcement. But the most troubling indication is the widespread view that government is both the other, the enemy, in existence not to recognize responsibilities but to impose arbitrary limits, and that it is itself a utilitarian commons, one of those resources that one uses as one can to maximize the satisfaction of one's wants.

C. Cost-Benefit Analysis

Whereas direct regulation has been the characteristic utilitarian response to disabling flaws in the working of markets, cost-benefit analysis has been its response to various "imperfections" in the democratic process. The major political "defect" is thought to be democratic processes' inability to reflect varying intensities in voter preferences. Every member of the electorate has but a single vote with which "to bid" for government projects and programs. Costs and benefits of those public initiatives, however, are dispersed very unevenly. In consequence, the democratic process is skewed toward endorsing programs that disperse net benefits to majorities of the electorate yet impose costs that in the aggregate exceed aggregate benefits. The "solution" has been to delegate project selection and program design powers to public officials who, independent of voter preferences, identify and add-up costs and benefits with an eye toward maximizing net "value".

In the United States, cost-benefit analysis was introduced during the 1950's and 1960's as the formal device for ranking public works projects such as dams, locks, and hydro-electric facilities according to relative "value". Analyses were used not so much to make decisions as to provide information to, and thus to influence, legislative authorizations and appropriations. Since then, the methodology has been employed with respect to a wide range of government project decisions, e.g., interstate highway, airport, and nuclear power plant siting, mass transit funding. It has been adopted as the primary mechanism for assessing all manner of proposed uses of the public domain, from offshore oil exploration and drilling to timber harvesting to construction of ski resorts. And from the 1970's, it has been employed as a mechanism for evaluating regulatory alternatives with respect to air and water pollution control, workplace safety, hazardous waste transport and dispersal, and so on. What is more, over this period the role of cost-benefit analysis has been transformed. Increasingly, it is used, not as a source of information for political decision-makers, but as the decision-making process itself. This transformation is partly the result of broad statutory delegations of power to administrative agencies, partly a matter of executive direction (a Reagan Executive Order directed all executive agencies to employ cost-benefit analysis as the criterion for decision-making where not statutorily foreclosed), and partly a matter of bureaucratic training and culture.

Cost-benefit analysis is commonly justified as a device designed to reach outcomes the market would yield if only it were able to operate, i.e., if monopoly, externalities, transaction costs, etc., were not so disabling. Thus there is the need to identify the consequences of government projects and programs, to characterize those effects as costs or benefits, to quantify them, and to price them. Whenever possible, market prices are used; and so the methodology tends to incorporate the same biases that the existing wealth distribution, monopoly and inability to internalize all costs and benefits give to market exchange. Also, costs and benefits recognized by market actors, viz., those that are certain, short-term, hard, and easily quantified, are readily taken into account in cost-benefit analysis, whereas those that ordinarily escape market notice, viz., costs and benefits that are uncertain, long-term, variable and cumulative in effect, are much more difficult to identify and quantify. Thus, outcomes tend to be skewed in favor of programs with "hard" front-end benefits and "soft" rear-end costs, e.g., timber harvesting, oil exploration and drilling, and against stringent regulatory standards with "hard" front-end costs and "soft" rear-end benefits, e.g. tight ambient air quality standards. This skewing of cost-benefit analysis outcome is of course greatly exacerbated by the use of high discount rates to reduce future costs and benefits to their "present values" (Reagan Administration agencies were told to use a discount rate of 10%!)

In addition, it is becoming increasingly apparent that use of cost-benefit analysis as a government decision-making methodology is afflicted with what might be called "scoping problems." One kind has to do with the scope of decisions that are made. Consider, for example, that cost-benefit analysis might be employed to select among alternative offshore oil exploration sites, but, because oil is treated by the market as a valuable commodity and so increased production is regarded as an unambiguous benefit, cost-benefit analysis cannot be used to decide whether a petroleum-intensive economy is better than some other. Likewise, while cost-benefit analysis might be used to select among alternative levels of air pollution control, because industrial output and displaced industrial workers are treated by the market as costs, cost-benefit analysis cannot be used to decide whether an industrial production oriented economy is what we "really want". In large measure, this scope-of-decisions problems is grounded in the cost-benefit analysis assumption that government decisions are no different than the decisions of businessmen within working economies. Altogether lost is the vision of government as shaper of "preferences", and "wants" and opportunities and conditions of life.

A second scoping problem has to do with the extensiveness of the cost-benefit analyses that are performed. Because gathering information is itself costly, every cost-benefit analysis at some point runs up against an inability to identify and so to consider indirect, long-term, cumulative, synergistic, cross-medial, interecosystem effects of alternative initiatives. Cost-benefit analysis discounts this informational scoping problem. It has no option, for treating it seriously would have the analyst overwhelmed with uncertainty. In consequence, though, we find air pollution control regulations generating water pollution problems, highway routing decisions with individually high benefit-cost ratios cumulatively generating congestion, and so on.

III. The Human within Structures of our Own Making

Although the conception of human as consumer continues as the dominant view expressed and effected by our law respecting environmental matters, there are indications that a third conception is coming into its own. Instead of disaggregating individuals into their activities, it seems to me that this third view moves in the opposite direction; it places individuals in contexts or structures within which they think, act, and live. There appear to be two main features to this move to structure. First, by putting the individual into context, he is seen to be far less self-defining than he is under the Enlightenment view and far less able to satisfy his preferences than he is under the utilitarian conception. What a person becomes and what consumptive activities he engages in are seen as circumscribed and greatly influenced by external conditions. Yet it seems to me that this third conception also insists that those external conditions that circumscribe what we can become and what we can do are largely products of human thought and action. We humans are seen as exercising the power, and thus also charged with the responsibility, to shape the conditions of our own existence. To put the point plainly: The state of the ozone layer influences whether we fry on the beach but we humans largely determine the state of the ozone layer.

Although this third conception of the human-in-context is in the process of developing, some of our existing law appears to manifest it. In this part of the paper, I offer some illustrations. I then speculate a bit as to the experience which seems to give rise to this third view and discuss some of its implications.

The Endangered Species Act, the Wilderness Act, and the development of the National Wildlife Refuge System are fairly obvious legal expressions of this third conception. In different ways, each of these initiatives has us establishing side-constraints on "normal" market and political outcomes. Also, each can be seen as shaping conditions under which otherwise unavailable preferences might be formed, e.g., to see some high mountain sheep, and under which otherwise unavailable paths of human development, e.g., medicinal use of rare plants, might be advanced. Costa Rica's series of national nature reserves is a similar effort with respect to rain forest ecosystems. And there are international analogues.

The development of technology forcing as a mode of pollution control is another manifestation. It has us exploiting market forces to generate the technical means to reduce emissions and effluents. Markets are seen not as impersonal mechanisms that coordinate satisfaction of inscrutable wants, but as structures that are created and directed--human artifacts like roads and bridges--as means to already identified ends. Further, as evidenced by the economic success of pollution control technology firms over the past decade or so, such initiatives belie the claims that environmental regulation is necessarily accompanied by economic decline and loss of national wealth.

The conception is also apparent, I think, in the fashioning of decision-making procedures, like the environmental impact statement process under NEPA and judicial employment of the public trust doctrine, that manifest a measure of caution and systematic inquiry into the uncertain consequences of "normal" market and political decisions and which are biased toward mitigation and preservation of "trust corpus". Short-term, "snapshot", aggregations of economic and political preferences are suspect. The decision-making focus is opened to the long-term, the cumulative, the not-easily quantified effects of human action. Those actions are thus viewed, not so much as produced by wants, but as generating conditions that then limit and foreclose otherwise available opportunities.

Interestingly, the sharp differentiation between the private and the public that is deeply imbedded in both Enlightenment and utilitarian thought appears to be breaking down. Each kind of governmental initiative manifesting this view of the human as shaper of his own conditions has its "private" analogues. Consider, for example, the success of efforts like The Nature Conservancy to raise funds for purchase of lands and other natural resources to be held off-limits to market and politically driven decisions on use. And consider the recent creation of mutual funds with portfolios restricted to stocks in "environmentally sensitive corporations", which, presumably, brings market pressures to bear in shaping the products and activities of those companies. And in many ways, individuals are beginning to see themselves as "trustees" of environmental conditions. Witness the grass roots recycling boom, the highway clean-up designations, the market demand for recycled products, and so on.

These developments, I think, rise out of a deep sense of regret that we have permitted ourselves to engage in a great deal of unnecessary destruction of our circumscribing physical conditions. The regret is not so much about our day-to-day activities. We know that private auto use stifles public transport, but we also recognize that that makes driving a car (and so adding a bit more to congestion and to hydrocarbon levels) more, not less, essential. Rather, the regret is that we have permitted conditions to become such that our opportunities are shaped in the ways that they are. So we push for the opening of recycling facilities, for reservation of wilderness areas, and for other structural changes that alter those opportunities.

This environmental ethic does not appear to be a "back to nature" move. Rather than rejecting technological advance and market forces, it uses them. It sees "back to nature" as back to ignorance. Nor, it seems to me, is it based on any notion of obligations to animals and plants and soils. We humans are much too anthropocentric for that. Instead, it dramatically broadens our conception of what it is to be human. It has us as consumers and as self-developing individuals, but also as trustees of the conditions within which we consume and develop. It rejects the sense of 'nature' according to which we distinguish the natural from the artificial. It says everything is artificial, everything is influenced by human hands.

International Aspects of Environmental Preservation

Jean Robert Leguey-Feilleux

Jean-Robert Leguey-Feilleux, Chair, Political Science Department, Saint Louis Univ., born and educated in France, received an MA in Business Administration and Economics at the Ecole Superieure de Commerce and Universite d'Aix-Marseille. He came to the United States as a Fulbright scholar and received an MA in Political Science from the Univ. of Florida. He earned a Certificate in English Culture and Institutions from the University of Cambridge, England and a Ph.D. from Georgetown Univ. in Political Science/International Relations. He was a Visiting Scholar at Harvard Law School and a Visiting Researcher at the United Nations in New York. He has authored articles and chapters in several volumes and co-authored a volume on The Law of Limited International Conflict. Listed in several Who's Whos, including Who's Who in the World, Leguey-Feilleux is a member of the Nat. Faculty of Humanities, Arts and Sciences. Twice recently, he spent time in Jordan, Israel, the Occupied West Bank and the Gaza Strip as a Malone Fellow.

Some environmental problems are exclusively local: local in scope, locally induced, and best managed at the local level, according to local preferences and means (certain aspects of urban decay fall in this category). However, much of today's environmental crisis is larger in scope. Our effluents do not respect political boundaries and national programs, however thoroughgoing, may not even begin to reduce transboundary pollution. Some countries in fact try to solve their domestic environmental problems by exporting them to less concerned nations thanks to the cooperation of irresponsible governments, as can be seen in the expanding international traffic in toxic wastes.

Preserving the environment on a global scale has become increasingly more complex. With the dismantling of colonial empires, international society is now divided into a record number of independent states (some 170 of them), all of them wanting to produce more, consume more, and as a result, applying greater and greater pressure on our ecosystems. The new nations are, of course, in a hurry to develop and industrialize, and are often willing to pay a high environmental price for fast, and seemingly less costly results. But we must admit that developed nations have only recently begun to show concern about the environment and their response has not been so much more responsible. [2]

It is now widely recognized that the world is facing an ecological crisis of unprecedented magnitude. A solution is perhaps not beyond our reach but we are hampered by the fact that our political outlook has not kept pace with the changing situation. It is still steeped in the 19th century, i.e., it is state-centered. Global society, in its political evolution, remains at the stage of the nation-state system. Much of the world tends to think in national terms and to have a very limited consciousness of global society or of the global common good. Nationalism remains an overriding factor in politics, and governments are all too quick to insist, in the name of sovereignty, that they may do what they please without regard for the needs of other nations. Thus, acute parochialism tends to color the policies of most nations. "The national interest," understood in insular terms is widely perceived as the ultimate justification of state action.

Our narrow priorities, therefore, tend to hamper the fulfillment of our moral responsibility toward global society. In fact, many refuse to accept this notion of "moral responsibility toward global society." We close our eyes to the rest of the world. Yet, our fate is increasingly dependent on what happens in other parts of this planet. This is particularly true in environmental matters since ecological systems transcend national boundaries. Nations became physically interdependent before they were prepared to accept the political consequences of this revolution. They still prefer to turn to national institutions, even when the tasks are beyond their means. To meet our many global problems, including environmental destruction, we must alter our political ethos.

> The earth is ultimately a common heritage, the fruits of which are for the benefit of all .... This has direct consequences for the problem at hand. It is manifestly unjust that a privileged few should continue to accumulate excess goods, squandering available resources, while masses of people are living in conditions of misery at the very lowest level of subsistence. Today the dramatic threat of ecological breakdown is teaching us the extent to which greed and selfishness -- both individual and collective -- are contrary to the order of creation, an order which is characterized by mutual interdependence.

> The concepts of an ordered universe and a common heritage both point to the necessity of a more internationally coordinated approach to the management of the earth's goods. [3]

We do not, however, have to start from scratch. The foundations of an international political order have been laid: Jolted by two world wars, governments have gropingly, and all too hesitantly, developed a network of international organizations now available to deal with environmental problems. [4] But there is no guarantee that their member-states are going to have the wisdom or the skill to use them effectively to do the job.

The UN remains a system of nation states, many of which continue to pursue, within the global organization, their own narrow interests parochially conceived. The clashes and power struggles characteristic of our post World War II era are of course taking place also in the UN. The behavior of states within its various organs is frequently an extension of their behavior without. Furthermore, when they are not clashing, member states are often satisfied with the appearance of cooperation however devoid of any substance. They formally endorse ambitious action plans but refuse to provide the means to implement them. [5]

These are serious limitations and the UN will not work miracles, unless one be tempted to call a miracle the fact that global cooperation occurred at all in the context of the bitter conflicts which have characterized international relations since 1945. Remarkably, in the margin of these many conflicts some useful work has been done without which the global ecological crisis would be much worse today. This is insufficient; but it provides a foundation on which more effective environmental preservation can be developed.

Involvement of the UN system in environmental work is often dated from the landmark conference held in Stockholm in 1972. This was indeed the first time that global society formally addressed the ecological crisis in comprehensive and systematic fashion. But it must be acknowledged that long before that, many UN agencies had been active in environmental protection activities. Some of these agencies, in fact, were created before the UN came into existence and established their UN affiliation after 1945.

The Food and Agriculture Organization (FAO) was founded by the conference on food and agriculture called by President Franklin D. Roosevelt at Hot Springs, Virginia, in May and June, 1943, and came into being in October 1945. Among its many duties were a large number of environment-related tasks such as land and water conservation, fertilizer utilization, protection against desert locust and other insects, fisheries protection and reforestation. New environmental tasks were added as ecological problems developed, for example, preservation of plant genetic resources. [6]

The World Bank (whose official name is International Bank for Reconstruction and Development or IBRD), was the end-product of another war-time conference convened by President Roosevelt in July 1944 at Bretton Woods, N.H. It came into existence in December 1945 with the initial task of post-war reconstruction, a mission with profound ecological implications in war-devastated areas. The Bank expanded its environmental protection efforts as new problems arose, with more funds made available to governments and international institutions for environmental preservation. [7]

The International Civil Aviation Organization (ICAO) was also initiated prior to the end of World War II, in the course of the Chicago conference of November-December 1944. Officially starting in April 1947, following twenty months of work by an interim organization to respond to the immediate post-war needs of civil aviation, this body is of course much more specialized and functionally limited than the preceding UN agencies; but it is concerned with a number of environmental issues related to air transportation such as meteorological data gathering and dissemination, airport ecology, physical conditions for safe and efficient air navigation, noise pollution and noise abatement, among other environmental concerns. [8]

Some agencies in the UN system go back even farther in time. The earliest among those concerned with environmental issues is the International Telecommunication Union (ITU), founded in 1865 (and now linked to the UN), an important function of which is to prevent chaos in the use of radio frequencies. This is a very confined environment, very crowded and easily disrupted if left unregulated. ITU apportions scarce radio frequencies among an increasing number of applicants. It is concerned with essential radio services such as maritime, air, and meteorological communications, with the relatively recent additions of space communications and radio astronomy. In particular, it coordinates telecommunication services for emergency situations, especially for rescue operations at sea. [9]

The International Labor Organization (ILO) is another pioneer, revolutionary in concept and structure, dating back to the Treaty of Versailles of 1919, and now integrated into the UN framework. It deserves recognition here for its important efforts for the protection of laborers in their work environment. The industrial revolution had a disastrous impact on this environment causing enormous human suffering and loss of life. Hundreds of ILO initiatives exposed flagrant abuses and generated remedial action. [10] An innovation in the structure of this organization insures the participation, with independent voting powers, of worker and management representatives (in equal numbers), in addition to the traditional governmental representation. This unusual tripartite structure remains in effect today. [11]

A number of other institutions of significance for environmental protection were established after the creation of the UN. The UN Educational Scientific, and Cultural Organization (UNESCO) was initiated in November 1945 (and in effect in November 1946). The constitution of this organization prescribed the preservation of the world's cultural heritage and led to projects designed to safeguard books, works of art, and monuments from a variety of natural disasters, the passing of time, accidents, or even the impact of national public works. The most spectacular task accomplished by UNESCO was the preservation of the monuments of Nubia (Operation Abu Simbel) from submersion by the waters of the Nile upon completion of the Aswan Dam in Egypt. Fostering deeper knowledge of our physical environment and facilitating interaction in this area among scientists and scientific organizations across national boundaries have been additional contributions. A good example of this kind of endeavor is the establishment under UNESCO auspices of the Intergovernmental Oceanographic Commission to spur international research on the oceans. [12]

The World Health Organization (WHO) was formally established in April 1948, after a conference, called by the UN, drew up its constitution in June 1946. WHO studies the impact of environmental factors on public health (housing, sanitation, economic and working conditions, and other aspects of environmental hygiene). It assists its member states in the provision of basic sanitary services, provides information and establishes criteria on the exposure of the population (or special subgroups) to biological and chemical agents or other physical environmental factors (e.g., sanitation), in the community and work environment. WHO draws appropriate standards for health protection; and it develops and coordinates programs for the monitoring of the levels, trends and effects of such exposure, and provides information on the protection of human health through the application of technology for the control of the environment. [13]

WHO's environmental activities are of special importance for the developing countries, where high mortality and morbidity result from biological pollution essentially associated with unsanitary environmental conditions (presence of insect or animal carriers of disease, lack of safe water supply, and unsanitary disposal of human and animal waste; problems are also related to food, housing and environmental conditions at the work place). National programs are often unable to meet present needs and further problems are created by the pace of population growth, and of industrial, agricultural and urban development. [14]

WHO efforts for the promotion of environmental health include: Technical advisory services to member states, particularly for the application of inexpensive but safe technology to solve health problems; stimulation of a flow of technical information from developed to developing countries and between developing countries; promotion of international agreements on environmental health criteria for the purpose of helping its member states in determining the health effects of various environmental hazards and planning abatement programs; assistance to member states (e.g., in the field of methodology) for effective environmental monitoring (e.g. sampling and measuring pollutants in the environment and generating internationally compatible information for a global assessment of trends); assistance to governments in the development and strengthening of institutions for the effective management of environmental health programs; training of interdisciplinary health personnel for the development of appropriate environmental health programs. [15]

The World Meteorological Organization (WMO), established in 1947, and in effect in March 1950, facilitates worldwide cooperation in the establishment of networks of stations for meteorological, hydrological and geophysical observations. WMO pays increasing attention to the monitoring of oceans and provides climatic information relative to every ocean. WMO fosters the establishment and maintenance of facilities for the rapid exchange of weather information worldwide, and promotes the standardization of meteorological observations and ensures uniform publication of relevant data and statistics. In addition, WMO studies the climatic impact of human settlements, energy uses, the growth and expansion of deserts, atmospheric and water pollution as well as the interchange of pollution between oceans and atmosphere; more generally, it is concerned with the interaction between human activity and the climate. [16]

Greater accuracy in weather prediction is high on the agenda of WMO, e.g., to predict the arrival of seasonal rainfall, upon which the economies of so many nations depend. Greater accuracy in forecasting tropical cyclones is an important part of its program in tropical meteorology. WMO research is also focused on weather modification (e.g., artificially induced rainfall), climate changes, and the interactions between atmospheric phenomena and other environmental factors (e.g., sea waves and ice and snow cover) .[17]

The International Maritime Organization (IMO) started as the Inter-Governmental Maritime Consultative Organization, whose statute was drawn up in 1948 by the UN Maritime Conference. It became operative in 1958. IMO deals with many aspects of the marine environment. It is concerned with maritime safety and the handling of dangerous cargoes and it is authorized to convene international conferences and to draft international maritime conventions such as the International Convention for the Prevention of Pollution of the Sea By Oil (amended in 1962), the International Convention Relating to Intervention On The High Seas In Case Of Oil Pollution Damage (1969), and the International Convention On Civil Liability For Oil Pollution Damage (1969). IMO establishes standards and guidelines for maritime activities (e.g., the International Maritime Dangerous Goods Code and the Manual on Oil Pollution). Its legal law committee tackles a wide range of problems of maritime which have a direct bearing on safety and pollution. Various questions of maritime pollution caused by ships are also handled by the Maritime Environment Protection Committee established in 1973.

The International Atomic Energy Agency (IAEA) was created in October 1956 and entered into effect in July 1957. As could be expected, this organization has important environmental duties to perform. It undertakes research on radioactivity and environmental contamination and establishes standards of safe practice including standards for labor conditions and standards of safety for the protection of health and the minimization of danger to life and property. It has identified the basic requirements for licensing and regulatory control of nuclear power plants and developed procedures for establishing limits for the release of radioactive material into the environment. [19]

IAEA issues regulations and technical guidance on specific types of operations and codes of practice and safety guides in the areas of nuclear power reactor siting, design, and operation. It provides similar guidance for the safe transport of radioactive materials, and gives emergency assistance to its member states in the event of radiation accidents. Radioactive wastes pose special problems. IAEA undertakes reasearch on treatment, safe management and disposal of radioactive wastes, and develops codes of practice in these areas. The agency contributes to the development of nuclear law; for example, it has grappled with the issue of liability in the event of nuclear accidents. This work has led to a number of international instruments, such as the Brussels Convention on the Liability of Operators of Nuclear Ships (1962), the Vienna Convention on Civil Liability for Nuclear Damage (1963), or the Convention Relating to Civil Liability in the Field of Maritime Carriage of Nuclear Material (1971). [20]

All of the international environmental protection work examined thus far is done by organizations falling into the category of UN Specialized Agencies. They are independent from UN General Assembly governance;

they have their own membership (separate from that of the UN), their own decision-making machinery, and they determine their own budgets. But these agencies are linked to the UN (by means of formal agreements), and they interact extensively with other UN bodies. Interagency cooperation is widespread and an elaborate system of coordination endeavors to prevent duplication of efforts. [21]

The Specialized Agencies, however, are not the only ones involved in environmental work. The UN General Assembly created a number of subsidiary organs which were led to address important environmental problems as a result of their primary missions (just as the Specialized Agencies did). These subsidiary organs are controlled by the General Assembly; their financing is included in the UN budget, but some of them are empowered to raise funds for program activities by means of voluntary contributions from UN members. They do not have the independence enjoyed by the Specialized Agencies but a number of them have, in practice, a good deal of autonomy in running their programs. The most prominent is the UN Children's Fund (UNICEF), a humanitarian institution born of the devastation of World War II (in 1946) and the urgent need to save millions of children from its aftermath. UNICEF soon turned to environmental work to protect the health of children, particularly in the developing countries. One of its largest projects in this area is a ten-year campaign in cooperation with the World Health Organization and other UN agencies, to improve environmental sanitation. Gastrointestinal and parasitic diseases are responsible for much sickness and many deaths among young children. To control them, UNICEF embarked upon an ambitious program of access to safe water and waste disposal. Large numbers of wells were dug and UNICEF spearheaded the development of simple, inexpensive and sturdy hand pumps for extensive distribution in the Third World, to bring people to abandon the use of contaminated surface water. This was supported by a program of education for better environmental hygiene and sanitation, with extensive community participation and special efforts to involve women in its implementation. By 1987, UNICEF had secured the cooperation of 95 countries for this environmental project. In 1986 alone, an additional 270 million people were given access to safe water and 180 million were given the benefit of effective waste disposal systems. [22]

Other projects are aimed at improving environmental conditions in urban slums, which have continued to grow in much of the developing world with dismal consequences for children, extending into adulthood.

UNICEF is also actively involved in the protection of children from the effects of armed conflicts and the devastation left in their wake. In 1986, some 40 countries were involved in some form of warfare; more than 70 percent of the victims were civilians, a majority of them women and children. [23] UNICEF cooperates with many organizations both public and private (such as the International Committee of the Red Cross, the Interamerican Health Organization, other UN agencies, and church organizations), in its activities in war-torn environments.

Relief work in devastated areas is also carried out by the Office of the (UNHCR) UN High Commissioner for Refugees and by the UN Disaster Relief Office (UNDRO). Refugee assistance, undertaken from its inception by the UN, has been particularly difficult because of the political sensitivity of the states involved. The General Assembly created the International Refugee Organization in December 1946, to cope with the problems caused by the Second World War; but its mission was viewed as temporary by the UN membership, and the organization was brought to an end in 1951. [24] It soon became clear, however, that the phenomenon would not disappear, and the General Assembly established the Office of the UN High Commissioner for Refugees. [25]

The preferred approaches to the refugee problem are repatriation or resettlement; but in many instances these are not feasible for all refugees (often for political reasons), and the UN is left with the task of dealing with the special environment of the refugee camps. Many private relief agencies participate in this humanitarian enterprise, particularly on the occasion of mass exodus. A separate refugee operation was initiated for Palestinians. The UN created in December 1948 a temporary agency (Relief for Palestinian Refugees) to supplement the emergency efforts of the International Committee of the Red Cross, the League of Red Cross Societies and the American Friends Services Committee. In December 1949, the General Assembly established a special organization better equipped to deal with the problem: The UN Relief and Works Agency (UNRWA) for Palestine Refugees. 41 years later, it is still providing assistance to more than 1.5 million Palestinians, many of them still living in camps in Jordan, Lebanon, Syria, Egypt, and Israeli occupied territories in the West Bank and the Gaza Strip. Some of the camps shelter more than 100,000 refugees. [26] More than 15,000 locally recruited UN personnel (virtually all of them refugees themselves) serve in the camps, providing sanitation, food rations, shelter, health services, education and vocational training (with the cooperation of the UN

Educational, Scientific and Cultural Organization). UNRWA depends entirely on voluntary contributions for the financing of its humanitarian mission, [27] and bitter political conflict in the whole region makes this work extremely difficult, as a consequence of which the camps remain utterly defective, and generate even more bitterness and radicalism.

Other environmental disasters (those resulting from natural causes), receive much international attention. Many components of the UN system are frequently called upon to provide assistance following earthquakes, floods, and other upheavals. As a result of a UN study of its involvement in these widespread relief and rehabilitation activities, the General Assembly created in 1971 the UN Disaster Relief Office. Its main task is to define specific needs, identify sources of assistance and coordinate the international effort. UNDRO's mandate includes also assisting governments in preventing disasters or reducing their effects by means of contingency planning. This is done in cooperation with a variety of voluntary organizations. [28]

In 1989, the General Assembly launched the International Decade for Natural Disasters, a 10-year effort (throughout the 1990s) to reduce the impact of these recurring calamities. This is to be achieved by means of a variety of projects. Among the early proposals are: Review of the lessons from previous disaster relief efforts; providing guidelines for hazard resistant construction techniques; development of an international earlywarning system for volcanic eruptions; improving telecommunications for disaster and emergency operations; and installation of an integrated international electronic information and communication network for disaster management. [29]

Specific disasters, especially those of long duration, have led to the establishment of special bodies, for example, the long-term drought affecting six countries immediately South of the Sahara, led to the opening of the UN Sahelian Office in 1973, to promote medium and long-term recovery. This office works in close cooperation with the Permanent Interstate Committee on Drought Control in the Sahel (known by its French acronym CICSS). The short-term relief effort was carried out by the Food and Agriculture Organization (as a result of a decision by the UN Economic and Social Council). Its relief work consisted in providing and transporting food rations, seeds, animal feed and vaccines. [30] In 1989, Soviet President Mikhail Gorbachev proposed establishing a program ("Green Cross") to provide emergency assistance to countries struck by environmental disasters. The UN agreed that the idea was a good one and gave the proposed Global Center for Environmental Emergencies priority consideration. It is not clear at this point, how this body would interact with UNDRO. [31]

Beyond relief and rehabilitation, environmental preservation has received much attention in the context of Third World development. The UN Specialized Agencies were among the first UN bodies involved in assistance to underdeveloped countries. In 1948, however, the General Assembly appropriated funds under its regular budget to enable the Secretary-General to provide teams of experts, offer fellowships and organize seminars to assist Third-World nations in their development projects, some of which pertained to environmental development. This was enlarged by the General Assembly in 1949 with the creation of the Expanded Program of Technical Assistance (UNEPTA), to receive voluntary contributions for this purpose.

In 1959, a special fund was created to increase the collection of voluntary contributions for Third-World development, and in 1965, the General Assembly decided to consolidate UNEPTA and the Special Fund and create a larger structure named the UN Development Program (UNDP) in order to raise still larger voluntary contributions and achieve greater coordination of all UN development activities by working with the various UN agencies already administering projects in this area. UNDP now generates more than $800 million annually and uses some of these funds to finance environmental projects administered either by other UN agencies (e.g., WHO and UNICEF), or by specific member states. [32] The World Bank is another source of funds for similar projects, along with the regional development banks (which are not part of the UN system). [33]

Population has been an important concern of the UN from the beginning of its existence. It was made the object of one of the functional commissions of the Economic and Social Council, and, in 1967, a UN Fund for Population Activities (UNFPA) was created by the General Assembly. It is now widely recognized that the rate of population growth has a major impact on the global ecology. In the 1990s, the world population will grow by 90 to 100 million per year. [34] By 2000, close to another billion will have been added to the current 5.3 billion. The largest increase will be in the poorest countries; but it must be noted that smaller population increases in the industrialized nations produce an enormous impact on the world's environment, given their enormous consumption and disastrously wasteful lifestyles.

The UNFPA has had a difficult role to play. Population issues tend to be controversial; they touch deep-rooted cultural and religious values and the Population Fund has had to navigate through these minefields making it prone to extreme caution. It has tried to distance itself from the abortion controversy by emphasizing time and again that it does not consider abortion to be a method of family planning; neither does it support coercive family control. It has been accused of being too cautious and of not doing enough; but it must be acknowledged that its funding (essentially by voluntary contributions) has for much of its existence been extremely limited (undoubtedly because of the controversial nature of the population question). [35]

The first World Population Conference called by the UN in 1974 showed enormous reluctance on the part of Third World nations to accept the notion of a global population problem emergency. But the dialogue generated by the UN produced many changes in outlook. By the early 1980s, many developing nations insisted that another world population conference was needed (it was held in 1984). In 1978, only 45 Third World governments considered their population growth too high; today, 67 governments, accounting for more than 85 percent of the developing world's population, have policies to reduce population growth. [36]

The global rate of growth has slowed down, but it remains alarmingly high. In 15 countries, 13 of them in Africa, birthrates rose between 1980 and 1985; in another 23 countries, the birthrate fell by less than 2 percent. Without sharp declines, the global population could grow to 14 billion during the next century (according to the 1990 report of the Executive Director of the Population Fund). [37] A better understanding of the relationship between rapid population growth and environmental degradation is likely to create greater interest in the issue. Another World Population Conference is planned for 1994; and, for the first time in its history, voluntary contributions to the Population Fund have reached $200 million. [38]

A different area of environmental concern has been addressed -- that of outer space. It did not take long to realize that uncontrolled national activity could easily damage spatial ecology. Early probes showed commendable efforts to prevent contamination of the solar environment;

but imperfect technology and the drive to achieve spectacular results before the competition did, opened the way to space degradation (witness the sowing of metal particles in earth orbit or experimental nuclear explosions in outer space).

In 1959, only two years after the beginning of outer space exploration, the UN General Assembly established a Committee on the Peaceful Uses of Outer Space which has been responsible for the development of a number of treaties to prevent a chaotic and destructive assault on this environment, but much more remains to be done in this area as greater and greater space activity is taking place. [39]

The UN Industrial Development Organization (UNIDO) created by the General Assembly in 1967, was given the task of speeding up industrialization in the developing nations. It was assigned the duty of coordinating all activities in the UN system concerning this endeavor. Obviously, its work has important ecological ramifications. In their rush to industrialize, many new nations are all too ready to ignore environmental considerations. In close cooperation with the UN Development Program and the World Bank, UNIDO is in a position to develop environmentally responsible industrialization plans. It is also a forum for negotiations between developed and developing nations in the area of industrial development. Its training program enables Third World personnel to learn more effective techniques in energy conservation, better utilization of nonrenewable resources and less environmentally destructive industrial methods. UNIDO became a Specialized Agency in 1985. [40]

Somewhat related to these concerns, the Commission on Transnational Corporations is now discussing hazardous technologies and safer alternatives and put forward a set of criteria to strengthen the participation of transnational corporations in environmental protection. The Commission is concerned about the role of these corporations in climate change in connection with development activities, and it proposed the creation of an environment fund financed by transnational corporations. At its Spring 1990 meeting, the Commission voted for a comprehensive study to be prepared for the 1992 UN Conference on Environment and Development, focusing primarily on how transnational corporations could provide access to and transfer environmentally sound technologies to developing countries on a concessional and preferential basis. [41]

The United Nations University (UNU) plays an important role in training people involved in development to be sensitive to environmental problems and it sponsors research on techniques appropriate to Third World levels of advancement for the purpose of achieving ecologically-responsible development. The UN University was created by the General Assembly in 1973. It has no campus and no faculty of its own; neither does it confer traditional degrees. Instead, it contracts with universities in various countries to provide the type of instruction or research meeting the needs of developing nations (ordinarily not provided by the institutions of the developed countries), e.g., short-term, highly specialized training programs for Third World professionals. It has generated important environmental research, for example for the development of better housing for the hot, arid regions, or for a better understanding of highland-lowland interaction, agroforestry development, or resource management in the humid tropical forests. [42]

Special mention must be made of the Economic and Social Council and the General Assembly of the UN. Both receive reports from their subsidiary organs, discuss issues of policy regarding the substance of their activities and mandate specific action (by these subsidiary organs), with environmental issues figuring prominently on their agenda. They can create additional structures, as needed, to attain their objectives. The Economic and Social Council submits recommendations for further action to the General Assembly, particularly as concerns budgetary matters and the convening of global conferences. The General Assembly is the main deliberative organ; its membership (160) is almost universal. Nations tend to be more attentive to its proceedings, and its decisions carry more weight. As a result, UN members often prefer to present major environmental issues in this forum. [43]

Other institutions foster cooperation at the regional level; they are generally concerned with a large range of issues among which environmental problems are now appearing more frequently. Five of these bodies are UN Regional Commissions. Others were independently created over the years. Among them can be cited the Organization of American States (OAS), the Organization of African Unity (OAU), the Commonwealth, the Council of Europe, the European Economic Community, the Nordic Council, and the Association of Southeast Asian Nations. [44]

Until the late 1960's, however, environmental work by individual organizations, important as it was in response to specific problems arising in connection with the fulfillment of their own missions, did not amount to a comprehensive environmental effort. It is true that consciousness of a global environmental problem was slow in developing. The turning point was in 1968, when the General Assembly, concerned about accelerating environmental deterioration, decided to convene a world conference on this issue. Sweden had a leading role in bringing this about: Its delegation in the Economic and Social Council and in the General Assembly prodded other delegations over a period of several months to achieve this result. Sweden hosted the conference in Stockholm in 1972. [45]

Wisely, four years were allowed by the General Assembly in preparation for this first comprehensive effort. A 27-nation preparatory committee was established to work out the details. A skillful administrator, Maurice Strong of Canada, was recruited, with the rank of Under Secretary General, in charge of the secretariat staff in Geneva, to prepare for the conference. Governments were requested to submit reports on their national environmental programs, and suggestions for international action. The UN Specialized Agencies were asked to report on their own past programs and their future plans for environmental protection. [46]

From the outset, Third World nations were extremely skeptical about the entire project. Current environmental destruction, as many of them saw it, was the result of rich-nation technological expansion. Industrialized nations had done precious little to preserve the environment as they acquired wealth and influence, and the developing nations saw themselves entitled to do the same or, if not going that far, they certainly did not want to see restrictions imposed upon their own industrialization programs. Rapid development was their preoccupation, not the environment.

In order to overcome this lack of interest or outright opposition, and induce developing nations to attend the Stockholm conference, regional seminars were organized in which officials from these nations could discuss the environmental situation and the purpose of the conference. Maurice Strong also initiated a conference of 27 experts to study the relationship between development and environment. In the end, most developing countries went to Stockholm and participated in the conference. [47]

This first comprehensive environmental venture attracted enormous popular attention. It was a media event of sorts. The Swedish Government made special efforts to enable representatives of private groups and nongovernmental organizations to follow the proceedings of the world conference. It organized a "Parallel Forum" or meeting for the benefit of nongovernmental organization representatives which was held simultaneously in another part of Stockholm. Almost 200 NGOs participated in its proceedings. [48]

The results of the 1972 Conference were far-reaching. Most important was global consciousness-raising and legitimization of environmental work, particularly among Third World nations. The conference did not completely eradicate their negative disposition. But participation in the proceedings convinced many that they, too, were threatened by environmental decay (in fact, that it could be a threat to their development), and that they did not have to repeat the mistakes of the industrialized world.

The conference produced a Declaration on the Human Environment, an Action Plan, and a recommendation to create a new UN agency exclusively concerned with environmental issues. [49] The General Assembly endorsed what the Conference had done, and created the proposed agency, calling it the UN Environment Program.

The Stockholm Conference was important in another respect: It had a profound catalytic effect. Influenced by the proceedings, many states embarked upon their own environment programs. In addition, many UN agencies incorporated environmental protection measures in the projects they were administering, or otherwise increased their environmental efforts.

The Stockholm conference was perceived as such a success that an entire series of world conferences came to be convened in the decade that followed, several of them on specific environmental issues, e.g., urban problems (1976), water pollution (1977), the destruction of arable land and the spread of deserts (1977), among others. [50] The UN has continued to use the world conference technique, but more sparingly for the development of international environmental law. [51] On the other hand, the members of the UN have refused to provide funding for a large environment program, and UNEP has remained an agency of comparatively modest size. Headquartered in Nairobi, Kenya (the first agency to be located in the Third World), it has a staff of 160 headed by an

Executive Director. [52] It maintains a small liaison office at UN Head-quarters in New York. Its board of directors, responsible for the development of agency policy, is the Governing Council composed of 58 UN member -- states elected by the General Assembly for three-year terms.[53]

UNEP has made up for its lack of financial resources by operating primarily as a catalyst, a facilitator, and an initiator of environmental projects (more than 1,000 since its establishment in 1973) financed directly by the states administering the projects, or by other UN agencies (such as the World Health Organization, UNICEF, the World Meteoro-logical Organization) in their own areas of concern and expertise. [54]

In addition, the UN Development Program, a funding agency with substantial resources (more than $800 million a year) allocates some of its funds to environment projects (or projects with environmental protection components). In 1989 it funded more than 400 projects with close to $300 million spent on specific environmental problems. [55] This is in addition to what many other UN agencies spend on their own initiative (and using their own resources) on environmental aspects of their programs. [56] It must be noted further that two important policy-and decision-making bodies of the United Nations (the Economic and Social Council and the General Assembly) address environmental issues in the course of just about every one of their annual sessions, either at the request of other UN organs or as a result of interventions by member states, and more projects are thus undertaken. [57]

UNEP's own programmatic contributions are substantial despite its limited means. Of great assistance is the gathering and disseminating of reliable environmental information. Scientists and environment profes-sionals around the world have all too often worked in isolation from their counterparts in other countries and used different means and standards of measurement, creating in the process much scientific uncertainty, confusing public opinion and hindering the development of appropriate responses. [58] An essential step in environmental management is therefore to determine with accuracy the extent of the problem and its causes, and then to make this information available to those who are to address the issue, such as policy-makers, as well as private or public agencies working in the area.

UNEP created <u>Earthwatch</u> for this purpose, with three main components:

1. The International Register of Potentially Toxic Chemicals used in particular for the regulation of import and export of dangerous substances. Also included in this part of the program is the task of maintaining a record of national chemical regulations of particular usefulness to Third World nations wanting to develop or improve their own control of such chemicals. [59]

2. The Global Environment Monitoring System (GEMS) designed to provide accurate data for environmental protection activities (includes also data on natural resources). In the Climate Monitoring Area of this program, for example, one network of 750 stations in 21 countries monitors the state of the world's glaciers whose advance and retreat are indicators of climatic change. This network is run by the World Meteorological Organization (a UN agency), UNESCO (UN Educational, Scientific and Cultural Organization), and the Swiss Federal Institute of Technology. Other networks of monitoring stations are used in different areas of environmental concern (e.g., health consequences of pollution). [60] As a result of this UNEP program, national governments have spent an estimated $200 million on environmental monitoring facilities. [61] One hundred and forty two nations have participated in at least one UNEP global environment monitoring activity and some 30,000 scientists and technicians worldwide have been involved. All of these now accept common measurement and instrumentation standards. [62]

3. The International Exchange of Environmental Information Program (INFOTERRA), created as a global network of national institutions designated by their governments to act as resources for this project. Each state facility thus becomes available to governments, industry and researchers around the globe. This network now includes 6,000 institutions possessing the needed information and some 600 commercial data-banks. INFOTERRA processes about 11,000 queries annually, both developed and developing nations making use of the network. Regional Service Centers are now in the process of being established to advise on common environmental problems.

In addition to the Earthwatch services, UNEP provides important legal assistance. International law is a potent tool of environmental protection; for that reason, UNEP established its own Environmental Law Unit to draft new international agreements, develop international guidelines, principles and standards, and give assistance to governments (particularly in the developing world) to strengthen their own environmental legislation and administration. UNEP publishes a Register of International-al Treaties listing 131 environmental agreements, twice the number concluded in the 50 years preceding the Stockholm conference. Over 115 countries have now established national environmental institutions (whereas only 25 had the benefit of such agencies prior to 1972). [64]

About one third of the new treaties have been negotiated under the auspices of UNEP's Environmental Law Unit, including the convention for the protection of the ozone layer. UNEP played a major role in the adoption of the Convention on International Trade in Endangered Species, the Convention on the Conservation of Migratory Species, and, most recently, the global convention on Control of Transboundary Movements of Hazardous Wastes (signed in March 1989). [65]

UNEP helps Third World nations in a wide variety of ways, for instance, by providing technical support in designing low-cost environment by monitoring systems, [66] or, by serving as a bridge between aid donors and recipients on environmental problems, through a clearing house program started in 1982 under which, by the late 1980s, about 100 projects in 50 countries were under way. For example, in 1983, the President of Botswana, a drought-stricken country, asked for UNEP's help in assessing his country's environment problems. A UNEP technical mission working with Botswanan experts developed a set of priority projects, eventually reviewed by a national panel and endorsed by the Government. UNEP then acted as a broker, working with the Government to raise funds from development agencies. By the end of 1986, nine projects were under way with funding from four countries, the UN Development Program, the European Common Market and the Government of Botswana. [67]

Other activities include both developed and underdeveloped countries, such as a long-range project to save regional seas from environmental destruction initiated by UNEP in 1974 and amounting to date to the development by means of negotiations with the governments concerned, of action plans for eleven maritime regions, among which are the Mediterranean, the Caribbean and the South Pacific Region, with over 120 coastal states participating. [68]

Thus, a good deal has been accomplished by UN agencies. But this work remains hampered by frequently hesitant or faltering member-state participation. The reluctance of nations to address global environmental issues is in fact striking, in light of mounting evidence of increasing environmental destruction. The international response to the ozone layer problem is characteristic of this phenomenon. It took ten years of scientific and diplomatic effort by UNEP to produce, in 1987, a treaty restricting the production and use of the chemicals causing much of the damage. Yet, after all this effort, governments could only agree to reduce their availability by 50 percent by the end of the century. UNEP was aiming at a total phaseout, but a majority of the states utterly rejected such a proposition. [69] This kind of obstruction, equivocation and propensity to settle for half measures is far from unusual. A new treaty was finally hammered out in London in June 1990, prescribing a total ban on most of the destructive chemicals by the year 2000 (with a ten-year grace period for the developing nations). But the person chairing the proceedings (UNEP's director) declared that the negotiations leading to the new agreement has been "the most taxing of his career."

The UN is even more hampered by inadequate funding. With two-thirds of its members poor or virtually bankrupt, the bulk of the financing must in practice come from industrialized nations. Most of these, with the United States in the lead, are adamantly opposed to any kind of increased spending. [71] It must be noted that most UN environment projects are supported by voluntary contributions, so that the financial problem has little to do with the controversy over the fairness of assessment rules. There are no rules in voluntary contributions -- affluent states are simply unwilling to contribute on a scale reasonably proportionate with the magnitude of the environment crisis. As a matter of fact, the United States has reduced its own contribution to UNEP. [72]

It is hard to determine accurately what the UN spends on environmental protection because a good deal of this work is done as an integral part of given projects in a wide variety of fields, [73] and it is hard to trace the amounts actually spent on their environment components. But we get an idea of how little the UN is allowed to spend when we know that the entire UN system (including the independent specialized agencies, and even the peace-keeping operations) spends just over $4 billion a year for all UN activities [74] -- international security, Third World development, world health, literally everything the UN does is included in this amount, and, of course, only a minute part of this is spent for environmental protection. UN resources are thus insignificant when judged against the needs of our planet. The United States has a budget that now exceeds one trillion dollars, and the world spends probably more than $800 billion a year on armaments. Where is our sense of proportion? Where is our sense of responsibility for the global common good?

Worse yet, a number of UN members are playing power politics with their financing, to the great detriment of substantive projects. The most remiss in this category today is the United States which, for years, has been withholding payment on its contributions, or threatening to do so, to keep the membership from taking measures it opposes. (The issues vary, but they are frequently tied to the conflict-between Israel and the PLO). The Soviet Government (prior to Gorbachev), and a small number of other countries have done it too at one time or another and on a smaller scale but this does not make the practice more tolerable. Currently, the United States owes more than $500 million to the UN.

UNEP, the only global agency working exclusively for environmental protection, receives only $5 million a year from the regular UN budget and approximately $30 million in voluntary annual contributions for its environment program. [75] In 1988, the United States spent $5 <u>billion</u> on its own Environmental Protection Agency. [76]

Resurging North-South conflict (particularly during the Fall '89 General Assembly session) over the global warming treaty and a projected global conference on the environment is adding a new wrinkle to global environmental efforts. Despite an overall tendency throughout the meeting of the General Assembly toward nondogmatic cooperation and pragmatism, a serious clash between environmental protection and development concerns achieved new prominence. [77] Third-World nations stress the continuing deterioration of their economic and social conditions, [78] and point out that poverty itself is a major cause of environmental decay, e.g., through deforestation, overgrazing, and loss of cultivable land through abusive practices. From their perspective, development and the eradication of poverty represent the best approach to environment problems and they advocate a massive effort of debt abatement and development assistance. They argue further that pollution of the atmosphere and other environmental damage are caused mainly by the high standard of living of developed countries and wasteful consumption of fossil fuels. [79]

There is of course a good deal of truth in this argumentation; but there is also an unmistakable element of power politics. The developed nations are anxious to achieve international cooperation on the problem of global warming. They want the UN to produce a treaty controlling emissions of the so-called greenhouse gasses -- and the price of Third-World cooperation is going to be more development assistance, western technology, debt relief, etc.

Another issue is involved, adding to North-South tensions. The members of the UN agreed that it was time to hold a second World Environment Conference and they chose to do it in 1992, the 20th anniversary of the Stockholm assembly. [80] Deciding what it will do is important and controversial. A major focus will be "environmentally responsible development," a concern rooted in the fact that poor nations, in their losing battle to improve their economic and social conditions frequently neglect the environmental consequences of their development policies and cause a great deal of environmental damage which ultimately hurts development. [81] Third World countries accept this diagnosis but they insist that to remedy the situation, they need the technological assistance of developed nations.

In the Fall of 1989, India, Sweden and a number of other countries proposed to the UN General Assembly the creation of an international fund, financed from taxes on pollution and voluntary contributions, to finance the purchase of environmentally sound technology by developing countries. Industrialized states however, were not prepared to make a commitment. Their position was that the 1992 conference should consider this proposal and any other ideas put forward. With regard to government assistance, the United States reiterated its argument that most environmentally safe technology (such as solar power equipment or scrubbers for smokestacks) belongs to private corporations and cannot be given away by states. North and South seem to be in a confrontation mode, and this is not going to help the 1992 conference or the global warming treaty. Initial negotiations for both have been difficult. [82] The creation of a similar fund, however, has been accepted as a part of the 1990 ozone arrangement. [83]

On the other hand, public awareness of environmental danger may now be at a turning point. A survey conducted by Louis Harris and Associates for UNEP in a sample of 14 nations (developed and underdeveloped) revealed a high level of popular alarm on the subject of environmental problems, the general perception being that the environment became worse in the last ten years. Interestingly, the survey showed that the level of concern is independent of the economic standing or education of respondents. In each country surveyed more than 75 percent of the people interviewed thought that stronger government action was needed, and a majority felt that international organizations like the UN should play a major role in addressing these problems. [84]

Governments seem to be responding to popular concerns. When the Governing Council of UNEP met in May 1989, a record 103 governments and 26 UN bodies sent representatives. The Council agreed to give top priority to treaties intended to control global warming and conserve biological diversity, and to increase UNEP's budget from $30 million to $100 million annually. Whether nations will actually provide the funds in another matter -- this program budget is funded on a voluntary basis. [85]

Pope John Paul II devoted his entire World Day of Peace message of January 1, 1990 to the environment crisis. It was the first papal document exclusively focused on ecology. The Pope condemned environmental abuse as "a seedbed for collective selfishness, disregard for others and dishonesty," and presented the crisis as a call to all people to ensure that creation be protected and preserved for future generations. [86] This urgent message is another sign of spreading environmental concern. Certainly, the prominence given to the environmental crisis in the mass media is helpful in creating greater popular awareness, but more needs to be done to emphasize the importance of international cooperation. If environmental protection is to be effective, it must be approached on a global level. We must outgrow our parochial tendencies and learn to keep our political differences from interfering with ecological endeavors.

We must realize also that it is difficult for us to enlist the support of nations less committed to environmental preservation if we refuse to work with them on issues which are of special significance to them. Greater responsiveness to other nations' needs ought to be viewed as indispensable to the attainment of our environmental goals. National self-centeredness is self-defeating. This is particularly true of relations with Third World nations. Their alienation is due, at least in part, to our neglect of their needs. Furthermore, we allow ideology and rhetoric to overshadow common interests.

Finally, global environment management requires more extensive UN programs. The machinery, by and large, is already in place; but the work cannot be done on the scale needed with the resources currently provided. International management is still in its infancy; member states are perennially worried about losing control and paying too much -- even when the amounts involved are minimal, as they are now. It may take some sort of cataclysm to produce more enlightened participation.

Our future is intimately related to that of global society. We should resolve to be more forthcoming in the task of fostering the global common good, more responsive to the needs of others, more active and more cooperative in international organizations, not for our own aggrandizement but for the sake of our common future -- a sound investment in ours too.

END NOTES

1.   Worldwatch Institute, State of the World, 1989, New York: W. W. Norton, 1989, Particularly, in this volume, Lester R. Brown, Christopher Flavin, and Sandra Postel, "A World at Risk," pp. 3-20. See also earlier Worldwatch Institute reports; and John Tessitore and Susan Woolfson (eds.), Issues Before the 45th General Assembly of the United Nations (an annual publication of the United Nations Association, USA), Lexington: D.C. Heath, 1990, pp. 83-117.

2.   See for example articles published in the annual reports of the Worldwatch Institute, State of the World.

3.   John Paul II, "Peace With All Creation," Origins, Vol. 19, No. 28 (December 14, 1989), paragraphs 8 and 9, p. 466. The full title of this World Day of Peace message was: "Peace With God the Creator, Peace With All of Creation." For a discussion of the enduring problem of national parochialism, see Richard,V; Sterling, Macropolitics, New York: Alfred A. Knopf, 1974, particularly, Introduction, pp. 3-21, and Part I, The Micropolitical World, pp. 23-323.

4.   See lnis L. Claude, Jr., Swords into Plowshares, 4th ed., New York: Random House, 1984, "Historical Backgrounds of Contemporary International Organizations,"pp. 21-80, which includes the genesis of the League of Nations and of the United Nations.

5.   The UN's perennial financial problems are an example of this phenomenon. A. LeRoy Bennett, International Organizations: Principles

and Issues, 4th ed., Englewood Cliffs, N.J.: Prentice-Hall, 1988, "Financial Problems," pp. 91-98. Evan Luard, The United Nations, New York: St. Martin's Press, 1985, chapter 6, "The Budgetary System: Finding the Money," pp. 113-136.

6.   See for example, UN Department of Public Information, Yearbook of the United Nations, 1985 (Vol. 39), Dordrecht: Martinus Nijhoff, 1989, pp. 287-1292. The International Fund for Agricultural Development (IFAD), another specialized agency, gives technical assistance grants to a number of research centers for environmentally significant projects. Among, the recipients are, the International Crops Research Institute for the Semi-Arid Tropics, Hyderabad, India, the International Center for Insect Physiology and Ecology, Nairobi, Kenya, and the International Irrigation Management Institute, Kandy, Sri Lanka, pp. 1365-1369, at p. 1367. In addition, see Issues Before the 45th General Assembly (1990), p. 86.

7.   1985 UN Yearbook (1989), pp. 1309-1313. Barber Conable, "Development and the Environment: A Global Balance," Finance and Development (A Quarterly Publication of the International Monetary Fund and the World Bank), Vol. XXVI, No. 4, (December 1989), pp. 2-4; and Jeremy Warford and Zeinab Partow, "Evolution of the World Bank's Environmental Policy, ibid., pp. 5-8; Leif Christoffersen, "Environmental Action Plans in Africa," ibid., p. 9.

8.   Evan T. Luard, International Agencies: The Emerging Framework of Interdependence (Royal Institute of International Affairs), Dobbs Ferry, N.Y.: Oceana, 1977, pp. 63-79.

9.   See International Telecommunication Union, Report on the Activities of the Union for 1984, Geneva: ITU, 1985. See also Luard, International Agencies, pp. 27-42.

10.   See David A. Morse, The Origin and Evolution of the ILO and its Role in the World Community, Ithaca, N.Y.: Cornell University Press, 1969. Also, 1985 UN Yearbook (1989), pp. 1278-1284; and UN Chronicle, Vol. XXVII, No. 3 (September-1990), p. 78.

11.   Robert W. Cox, "ILO: Limited Monarchy," in Robert W. Cox and Harold K. Jacobson, and others, The Anatomy of Influence: Decision Making in International Organization, New Haven: Yale University Press, 1974, pp 102-138.

12.   1985 UN Yearbook (1989), pp. 1295-1299; Luard, International Agencies, pp. 186-189.

13.   WHO: "What it is; What it Does, Geneva: World Health Organization, 1988.

14.   The Work of WHO, 1986-1987, Biennial Report of the Director-General, Geneva: World Health Organization, 1988, particularly, Chapter 12, "Promotion of Environmental Health," pp. 125-135.

15.   Four Decades of Achievement, 1948-1988, Highlights of the Work of WHO, Geneva: World Health Organization, 1988. See also, 1985 UN Yearbook, pp. 1302-1305, and 773-787.

16.   1985 UN Yearbook (1989), pp. 1350-1353.

17.   "The major effort within WMO on atmospheric ozone resided in its Global Ozone Research and Monitoring Project. The project aimed at improving the worldwide ground-based data collecting network for ozone (the Global Ozone Observing System), studying the distribution and changes of non-CO_2 "greenhouse" gases and participating in the ongoing efforts to assess ozone changes. Atmospheric CO_2 was monitored by WMO under a separate programme. WMO coordinated, organized or otherwise participated in international assessments of the state of the ozone layer in cooperation with other international and national agencies." 1985 UN Yearbook (1989), p. 1352.

18.   1985 UN Yearbook (1989), pp. 1356-1357; see also Luard, International Agencies, pp. 44-61.

19.   1985 UN Yearbook (1989), pp. 1271-1275, and Lawrence Scheinman, "IAEA: Atomic Condominium?" in Cox, Anatomy of Influence, pp. 216-262, particularly at pp. 236-238, and 230-234.

20.   1985 UN Yearbook (1989), pp. 1272-1273; see also pp. 693-696, and Luard International Agencies, pp. 120-131.

21.   Luard, International Agencies, Chapter 16, "The Co-ordination of International Government, pp. 264-286; also, 1985 UN Yearbook (1989), pp. 831-833, 1038-1044.

22.   UNICEF, Rapport Annuel, 1987, New York: UNICEF 1988, p. 21.

124

23.   UNICEF, Rapport Annuel, 1987, pp. 31. See also <u>1985 UN Yearbook</u> (1989), Children, Youth, pp. 962-982.

24.   Bennett, 3d ed., pp. 287-289.

25.   <u>Ibid</u>., pp. 288-289; it was established in 1951. See also <u>1985 UN Yearbook</u> (1989), Chapter XXI, "Refugees and Displaced Persons," pp. 990-1011.

26.   The author visited several of these refugee camps in 1988 and 1990 in Jordan, the Israeli-occupied West Bank, and the Gaza Strip.

27.   Bennett, 3rd ed., pp. 289-290. UNRWA, <u>A Survey of United Nations Assistance to Palestine Refugees, UN Relief and Works for Ag Agency for Palestine Refugees,</u> 1972.

28.   Harold K. Jacobson, <u>Networks of Interdependence</u>, 2nd ed. New York: Knopf, 1984, pp 342-343. UNA-USA Policy Studies on International Disaster Relief, <u>Acts of Nature, Acts of Man: The Global Response to Natural Disasters,</u> New York: UN Association, USA, 1977.

29.   <u>Issues Before the 45th UN General Assembly (1990)</u>, p. 110.

30.   <u>Issues Before the 43rd UN General Assembly (1988)</u>, pp. 28-30.

31.   <u>Issues Before the 45th UN General Assembly (1990),</u> p. 108.

32.   Robert E. Riggs and Jack C. Plano, <u>The United Nations,</u> Chicago: Dorsey, 1988, pp. 331-335; also, Bennett, 3rd ed., pp. 235-238.

33.   Riggs and Plano, pp. 335-340.

34.   Bennett, 3rd ed., pp. 270-271; <u>Issues Before the 45th UN General Assembly (1990).</u> pp. 92-93; see also <u>UN</u> Chronicle, Vol. XXVII, No. 3 (September 1990), p. 79.

35.   Bennett, 3rd ed., p. 271. <u>1985 UN Yearbook</u> (1989), Chapter XIV, "Population," pp. 760-771.

36.   <u>Issues Before the 45th UN General Assembly, (1990)</u> p. 95. See A. W. Clausen, <u>Population Growth and Economic and Social Develop-</u>

ment, Addresses by the President of the World Bank and the International Finance Corporation, Washington, D.C.: World Bank, 1984.

37.   Issues Before the 45th UN General Assembly (1990), p. 93.

38.   Ibid., p. 100; see also Issues Before the 44th UN General Assembly (1989) pp. 108-115, and Issues Before the 43rd UN General Assembly (1988), pp. 105-111.

39.   Luard, International Agencies, pp. 80-91, and see UN Chronicle, Vol. XXVII, No. 2 (June 1990), pp. 35-37.

40.   Riggs and Plano, pp. 330-331. Issues Before the 43rd UN General Assembly (1989), pp. 185-186. 1985 UN Yearbook (1989), pp. 591-615.

41.   UN Chronicle, Vol. XXVII, No. 3 (September 1990), p. 59.

42.   United Nations University. Tokyo: UNU, 1985. The United Nations University: The Fifth Year, 1979-1980, Tokyo: UNU, 1980; see also 1985 UN Yearbook (1989), pp. 785-786; also pp. 687-688.

43.   Lawrence S. Finkelstein (ed.), Politics in the United Nations System, Durham, N.C.: Duke University Press, 1988, pp. 217-220.

44.   Bennett, 3rd ed., "United Nations Regional Commissions," pp. 380-382; "Multipurpose Regional Organizations," pp. 354-365; "Regional Development Banks" pp. 378-380. See also Robert W. Gregg, "The UN Regional Economic Commissions and Multinational Cooperation," in Robert S. Jordan (ed.), Multinational Cooperation: Economic, Social, and Scientific Development, New York: Oxford University Press, 1972, pp. 50-109.

45.   Bennett, 3rd ed., p. 295.

46.   Ibid., p. 296.

47.   Ibid.

48.   Ibid., p. 297.

126

49.   Ibid., pp. 297-298.

49a.   Nicholas A. Robinson, prepared statement, Hearings Before the Subcommittee on Human Rights and International Organizations of the Committee on Foreign Affairs, House of Representatives, Ninety-Seventh Congress, 2nd Session (March 30, April 1 and 2, 1982), Review of the Global Environment 10 Years After Stockholm, Washington: U.S. Government Printing Office, 1952, pp. 361-366.

50.   Bennett, 4th ed., "The World Conference Technique: A Developing Trend," pp. 293-324.

51.   See examples on p. 24.

52.   By contrast, the World Health Organization has a staff of about 4,500. UN Yearbook, 1984, p. 1224. See also, U.S. Department of State, United States Participation in the UN, Report by the President to Congress for the Year 1986, Washington: Government Printing Office, 1987, "United Nations Environment Program," pp. 154 ff.

53.   See Annual Report of the Executive Director, 1986, Parts I and II, Nairobi, Kenya: United Nations Environment Programme, 1987. The Governing Council is composed of 16 African, 13 Asian, 10 Latin American, 13 Western European and Other, including the U.S., and 6 Eastern European nations. It now meets biennially. US Participation in the UN, 1986, p. 155.

54.   UNEP Profile, Nairobi, Kenya: UNEP, 1987, pp. 2, 30, 35. Some of these agencies run programs with important environmental components (independent of projects initiated by UNEP). For example, the International Atomic Energy Agency works in the area of radiological safety and radioactive wastes; the UN University is concerned with problems of resource management and of fragile ecology, to mention a few of them.

55.   John Tessitore and Susan Woolfson, eds. Issues Before the 44th General Assembly of the United Nations, United Nations Association of the USA, Lexington: D.C. Heath, 1990, p. 122.

56.   There are about 19 independent agencies and some 30 other bodies initiating and administering programs. See examples in note 54.

57.   See for example Issues Before the 44th General Assembly, pp. 119-127.

58.   UNEP Profile, p. 24.

59.   U. S. Participation in the UN, p. 156.

60.   GEMS concentrates on five areas of environmental concern: Climate, transboundary pollution, terrestrial renewable natural resources, oceans, and the health consequences of pollution. UNEP Annual Profile, p. 24. See also, for example, UNEP Annual Report (1986), Part I, pp. 50-54.

61.   UNEP Profile, p. 30.

62.   Ibid., p. 24.

63.   Ibid., p. 26. See also UNEP Annual Report, 1986, Part I, pp. 55-58.

64.   UNEP Profile, p. 27.

65.   Issues Before the 44th General Assembly, pp. 126-127.

66.   U.S. Participation in the UN, p. 156.

67.   The four countries were: The Netherlands, Norway, Sweden, and Britain. Among the funded projects were: plans for improving the firewood supply; research on range management and restoration; and the reorganization of water administration. UNEP Profile, p. 31.

68.   In the last-mentioned region, in 1986, the United States joined with six other governments in signing the Convention for the Protection of the Natural Resources and Environment of the South Pacific. This was the outcome of three years of negotiations funded in part by UNEP. U.S. Participation in the UN, 1986, p. 156. With regard to the Mediterranean, see "The Mediterranean in Action," The Siren, News from UNEP's Oceans and Coastal Areas Program (December 1987), pp. 1-2.

For a discussion of UNEP's other projects, see for instance, UNEP, Proceedings of the Governing Council at its Fourteenth Session, Nairobi,

June 8-19, 1987 (UNEP/GC.14/26 July 10, 1987), pp. 48-78, and for future directions, UNEP, Proceedings of the Governing Council at its First Special Session, Nairobi, March 14-18, 1988, "System-Wide Medium Term Environment Programme in the Medium Term Plan of the United Nations Environment Programme for the Period 1990-1995," pp. 11-31. (UNEP/GCSS. I/7 March 28, 1988).

69.   Issues Before the 44th General Assembly, pp. 125-126; see also Issues Before the 43rd General Assembly, United Nations Association of the U.S., Lexington, Mass.: D.C. Heath, 1989, pp. 119-120.

70.   New York Times, June 30, 1990, p. 2. And one easily forgets how long the issue of environmental destruction has been debated. See for example Roy L. Meek and John A. Straayer, eds., The Politics of Neglect: The Environmental Crisis, New York: Houghton Mifflin, 1971. See also House Committee on Foreign Affairs Hearings (1982), "The United Nations and the Environment: After Three Decades of Concern, Progress Is Still Slow," by Patricia Scharlin, pp. 378-381.

71, Ibid., Testimony by Erik Eckholm, at p. 338. The Soviet Union has, for years, been in complete agreement with the United States on this.

72.   Pulling Together: A Program for America in the United Nations. Final Report. New York: United Nations Association of the U.S., 1988, p. 35.

73.   Environmental protection thus receives more than formally tabulated under that heading; but compared with the magnitude of the task, this amount remains inadequate.

74.   Caution is needed in using UN agency figures: Many UN bodies have biennial budgets. Each specialized agency has its own budget. These figures are therefore not included in the budget voted by the General Assembly. See Bennett, International Organizations, pp. 91-98; and Robert E. Riggs and Jack C. Plano, The United Nations, Chicago: Dorsey, 1988, pp. 48-55.

75. By contrast, the World Health Organization has an annually adjusted budget of $250 million. UN Yearbook, 1985, p. 1305. See UNEP Profile, p. 30, and U.S. Participation in the UN, 1986, p. 155.

76.   UNA-USA, Program for America in the UN. p. 35.

77.   This conflict had been prevalent prior 1972. New York Times, December-20, 1989, p. 3.

78.   The record of the decade just ended indicates a decrease of Third World living standards of one percent a year on average. New York Times, November 5, 1989, section 4, p. 2.

79.   New York Times, November 9, 1989, p. 4.

80.   New York Times, January 3, 1990, p. 5. See also Issues Before the 44th General Assembly, p. 122.

81.   The World Commission on Environment and Development established by the UN General Assembly at the behest of UNEP in 1983 and headed by Gro Harlem Brundtland, (Norwegian Environment Minister, later Prime Minister) and composed of 22 highlevel government decision-makers, scientists, economists, lawyers and educators from different nations, studied the environmental destruction wrought by development and discussed it in its final report (the Brundtland Report entitled Our Common Future) which was submitted to the 42d General Assembly in 1987. Issues Before the 44th General Assembly, pp. 120-121.

82.   New York Times, January 3, 1990, p. 5. Adding to the difficulty, the UN is concurrently negotiating its Action Plan for the Fourth Development Decade. These negotiations are to continue during 1990; as anticipated they are generating a good deal of North-South friction. See New York Times, November 5, 1989, section 4, p. 2.

83.   New York Times, June 30, 1990, p. 1, 2.

84.   The 14 countries were: Norway, West Germany, Hungary, Senegal, Nigeria, Kenya, Zimbabwe, Saudi Arabia, India, China, Japan, Argentina, Jamaica, and Mexico. UNEP-Harris Survey, May 9, 1989. Issues Before the 44th General Assembly, p. 120.

85.   Ibid., pp. 123-124. The Council adopted an eight-point program of activities: Protection of the atmosphere by combating climate change and global warming, depletion of the ozone layer, and transboundary air pollution; protection of the quality of fresh water resources; protection of

ocean and coastal areas and resources; protection of land resources by combating deforestation and desertification; conservation of biological diversity; environmentally sound management of biotechnology; environmentally sound management of hazardous wastes and toxic chemicals; and protection of human health and quality of life, especially the living and working environment of poor people, from environmental degradation.

86.   John Paul II, "Peace With All Creation," Origins, Vol. 19, No. 28 (December 14, 1989). pp. 465, 467-68. The full title of this World Day of Peace message was "Peace With God the Creator, Peace With All of Creation." Repeatedly, the Pope stressed that the ecological crisis is a moral crisis.

GENERAL DISCUSSION

SESSION I

ABELL   I am not going to re-state my paper. I'll mention a few of the things in the paper along with a brief treatment of a few things that were not in the paper. In the paper I talked about climatic change but not about precipitation. I stayed mainly with warming and cooling. This business about the greenhouse effect -- and things that are being said about it -- has been bugging me for a long time. My first professional position was in the Office of Climatology. During my stay there I was working almost exclusively on drought; my research was on the Western Great Plains. I worked next door to J. Murray Mitchelle who was one on the early investigators and proponents of a carbon dioxide (CO_2) greenhouse warming. Murray did a very good job. In my opinion he documented very well the warming in the United States and North America through the first 50 years of the century. He was very meticulous in his selection of weather stations, climatological stations. He was very careful to select stations where the environment around the station was best to discern net change. That's the first thing I'd like to talk about.

We have to be very careful in using some of the recent work that's being done. Not only have station sites been moved, but the environments around the stations have changed. A number of us here are from St. Louis, so I might use St. Louis as an example. When we talk about a climatic normal or a climatic average -- this is National Weather Service wide -- we use the latest 30 year averages. We do it every ten years. If I had a 30 year average on temperature where I incorporated the 1930s, the '40s, and the '50s, I submit to you I would be talking about temperatures much warmer and precipitation values much lower than if I were to have used 30 years prior to that or 30 years since then.

As far as the station itself is concerned, St. Louis would be a very bad data point to use. Up to the 1930s, observations in St. Louis were made close to downtown. There were several moves, but the station effectively remained in the city. In the late 1930s the station was moved to Lambert Field. Originally it was on the west end of Lambert. There was an old terminal building that the National Weather Service shared with the FAA. When they moved out there, it was a different ball game. They moved out to a real suburban environment as opposed to an urban environment. Immediately we started seeing cooler temperatures, particularly on still nights (not much wind) and on nights without a lot of cloud cover. Cloud cover and/or wind is a great temperature equalizer overnight. Then the Weather Service had to move again. When a runway had to be expanded, the station was moved to the other side of Lambert near McDonnell Boulevard; immediately the temperatures went up 5 degrees. We couldn't see the reason for it, but the temperatures on the average on those clear, still nights went up 5 degrees. I've been forecasting for the media regularly since 1972 and on those still nights I give people a city forecast and a suburban forecast (a country forecast) as far as temperatures are concerned. For the last three or four years I've used almost the same minimum temperature for Lambert Field as for the city of St. Louis because the area around the airport has grown up so much.

The Weather Service didn't want their forecasters nor their data readers outside in the elements, so they came up with a new automated weather observation system where, in effect, the forecaster could stay inside the office. For some strange reason, they built it over concrete. This happened within the last six months, and the situation got worse. After some criticism, some from the National Weather Service itself, they decided to move it back over the grassy area. When we start looking at things like that, St. Louis would be a bad selection. In some of the studies I've seen, they've been careless in their selection. We have to be cognizant of that when we read some of these things.

I mentioned "feedback" in my paper. It's extremely complex. Let me review that very quickly. One climatic element will interact with another climatic element. The atmosphere will interact with the hydrosphere (ice and/or liquid water) on the earth. I gave you the example where cooling and increased precipitation would lead to increased glaciation. That increases the albedo and leads to further cooling and perhaps more glaciation. That's a positive feedback effect. Then take the example of CO_2. If we do get a warming -- it's theoretically sound, but how much warming we might get we don't know -- we should get more evaporation. If we get evaporation, we're dealing with another greenhouse gas, water vapor. There's a lot more water vapor than CO_2 in the atmosphere. People look at that and say, "oh, runaway warming." By the same token, however, more water vapor might increase the cloudiness. If we get more cloudiness and precipitation, we're going to get more albedo. If we were to get a lot of snow, which may eventually go into glacial ice and perhaps encourage the growth of glaciers, we would get a great increase in the albedo. That would then give us a cooling which would counteract the CO_2 warming. There's a lot in this that we really don't know. We think we know what the trends are, but it's a question of how severe a warming are we experiencing.

I brought up an example of anthropogenic fire in terms of changing a landscape and I purposely used the example of the native Americans because they did have a great respect for nature. People talk about the pristine environment; it was more pristine then, but these peoples had their way of doing things and changed the landscape on a very regular basis, especially by burning. In the 1850s the smoke from the fall burning was severe in the Pacific northwest year in and year out. There was serious consideration given to putting lighthouses on the Columbia River so that navigation could continue. Not only did these people do something that perhaps changed the face of part of this country, but there were health considerations involved as well.

Cooling or warming impacts on precipitation because the weather belts move. The storm tracking jet stream will move equator-ward with cooling; it will move pole-ward with warming. We're looking at changing our temperature contrast between land and water. We're changing the monsoonal circulations, increasing them or decreasing them. We can certainly do it on a small scale. We had a study here in St. Louis in the 1970s, the Metromax Project, where we, in effect, showed that we had two precipitation maxima within the St. Louis area, one in Edwardsville and the other in Centerville, both in Illinois. Initially we felt the reason was that these areas were downstream from the St. Louis industrial air pollution plume. I think that that's a part of it. I'm required weekly to start or initiate studies of cloudiness which eventually may grow into precipitation. Industry and its pollutants would certainly account for a part of those precipitation maxima. But we've come to realize that, as well as the addition of these particles to the atmosphere, the thermodynamic effect (the heat plume) of the St. Louis area itself is probably having as much, if not more, of an effect on that precipitation maximum. We're not only getting more precipitation there, we're getting more lightning. Although we're not getting enough hail storms really to make a good study on them, but we seem to be getting more hail as well.

I didn't treat acid rain in my paper. That's a real problem, too. I was delighted to see in one of the papers NO_x coming in for part of the blame on this, as opposed to just SO_x, the sulfur oxides. Some of the NO_x, of course, comes from the pretty little blue flame, the hot burning natural gas, and some comes from automobiles with the nitrogen oxide emissions as well.

We have to be very careful with the greenhouse effect as well as acid rain. If you watch television or if you pick up a newspaper, you're going to see article after article about these things. Most of them, I think, come from people who don't seem to know anything about it. The media feeds on things like this, things that have a veneer of scientific information. It's like a feeding frenzy. We saw the cold fusion fiasco in Utah. In the St. Louis area in the last six months, we've had the dynamic duo of Iben Browning and David Stewart talking about something terrible that's supposed to happen on December 3 [a forecast of a 50/50 chance of a 6+ earthquake along the New Madrid Fault in southeastern Missouri]. If it happens on December 3, my attitude will be the same -- it was just a roll of the dice. The same is true of the greenhouse warming. It's been going on for years; but it's accelerated in recent years. The point I hope I've made is that we do have a natural climatic change. Also we're certainly having an impact. We think we know how, but we don't know how much. Some of the remedies that we've seen proposed are a little scary. I think many of the proposals are quite premature. Thank you.

BENNETT   The substance of my paper was people's awareness of environmental issues. I made a point that humankind has always been in a precarious balance with nature and the things we've done have always impacted the environment. I started with the hunting-gathering peoples. They've been around for 100,000 years. This naked ape that was wandering through the savannas and exploring the world was basically following the food supply. If they had stayed in one place, they would have starved; but they were smart enough to follow the animals. Therefore, the environment didn't degrade from overgrazing or from deforestation for fire wood or shelter. The human waste which was left behind, garbage and sewage, was allowed to decompose naturally, so it never occurred to anyone that it could cause environmental problems.

As people became more civilized, these problems didn't vanish. They weren't just out of sight, out of mind -- they were there every day. But humans were often slow to learn the lessons. We know where a lot of these early civilizations were. We find pottery, tools and other evidence of a horticulture, but we don't know who they were. They obviously were intelligent. They were rising above nature, controlling nature, and producing their own crops. But for a variety of reasons, they didn't survive. Either it was mismanagement of water or soil erosion or sometimes plagues. Throughout the Middle Ages, Europe was ravaged by plagues. Much of the reason for that was their inability to take care of their sewage. Even when they did learn, they were slow to respond. Evidence of that comes up into our own times. In the 1880s, when the Europeans who had settled the continent were driving the Indians onto reservations, they made no provisions for proper sewage. The Indians themselves didn't know how to take care of it. They had always moved away from it, as their ancestors did. Once they were forced to live in tight population circumstances, mismanagement of sewage led to epidemics that ravaged the Indians. The European-Americans saw this as evidence of the low quality of the Indians' intellect. Not much was done about it. Even now, much of the education on native American reservations does not cover basic health issues. It's also a problem in the agricultural fields of California. Migrant farm workers are not given good sanitation facilities. A lot of illness occurs from that lack of basic education.

In solid-waste issues the media grab a story and just run with it. If we find a toxic waste dump or something like Times Beach (a small town in Missouri that was contaminated with dioxin-laden waste oil), people always want to say: "Here's the problem. It's their (whoever they are) fault. They did it." But what we're finding are just examples of the larger problem. As life is changing, we're not learning fast enough. As waste problems caused health problems in the Middle Ages, so too are they causing us problems. Thus, in my paper, I suggested that modern populations (Americans, in my work) get more involved in learning about the problems, so that, when they hear something about a waste dump, they realize that it's not a single individual who was so mean spirited that he or she wanted to dump something in somebody else's water supply.

I wrote about some of the problems in St. Louis. There's an area of former wetlands north of downtown that was a convenient place to dump trash. Even though St. Louisans had learned about earlier mistakes in waste management, they took all their garbage outside of the city limits and dumped it in the wetlands. All sorts of stuff is buried there. It's industrialized now and some people have been mismanaging hazardous materials there. If you take a soil sample, there's no telling what you'll find. We have a lot of sites there that were proposed Superfund sites. We can make the company clean up what can be attributed to them. After they've done the cleanup, the soil is often more contaminated than before. What was underneath the individual company's dump site was even worse, or worse in other ways. We can't deal with that kind of a situation with the typical reaction: "It's their industry; they made the money, it's their fault." It's everybody's fault because we didn't learn fast enough. Dealing with it will demand more cooperation with more groups of people involved.

I then pointed out in my essay how people could learn more about these issues. The dearth of students in the sciences for the last 20 years has been attributed to our sliding industrial output. Educators, learned societies and governmental bodies are trying to get more people interested in biology, chemistry and engineering. I think we could develop student interest in science through ecological studies. By making the hard sciences more clearly applicable, we might bring the sciences alive for them, teach them about problems, and, in the process, become better citizens in making judgments on things like the global warming. This is something nobody thought about for years. An article would come out here and there. When it made the cover of *Time*, everybody got concerned. Now people are talking about things like carbon meters and carbon taxes, but they still don't really know what the situation is.

I saw a debate between two professors on global warming theory, and the only thing they really agreed upon was that the EPA model for evaluating it was totally inadequate. What's happening and how fast it's happening is something we don't know. More information will come in. Let's hope we can figure it out before disastrous circumstances come about.

Aside from teaching, I suggest that interested adults look at other ways to learn about the issues. I've tried to do that giving people the chance to discuss various problem areas. Many people read the paper and see only the headlines. There are others, however, who are far more experienced in talking about these issues, in raising new questions and developing new sources of information. Then, if there is going to be a bill passed or something like that, they can work with the politicians to get a more cohesive policy. That is something I noticed in dealing with regulations. When an issue comes up, politicians want the headlines and a bill is brought through. Often it's not followed up on. An unfortunate example of that is Pat Dougherty's [member of the Missouri legislature] solid waste bill. For people from outside the state who don't know about this bill, it is a comprehensive solid waste management bill that will keep tires, batteries and all such waste out of the landfills. It's also designed to promote recycling. He got it through the legislature and most observers say it's a good bill. But he forgot to provide for fulltime employees to implement the plan. There's no money available to the Department of Natural Resources for almost a year to hire anybody to implement this law. So now we need emergency funding for fulltime employees to get this solid waste program on the right track.

Emergency funding is difficult to get from the legislature. Moreover, the legislative and executive branches don't always see eye to eye on such things. The Department of Natural Resources has always had a lot of trouble with funding. Recently it almost lost the Water Pollution Control Program because of lack of cooperation among branches of government. If we set up a program without the proper implementation, it will be difficult to achieve the legislative goal. Perhaps Pat Dougherty didn't put any funding requests in it because he wanted the bill to pass. If people were better informed, he might have had the support to get the money to do it right in the first place.

I also made the point that there are a lot of the unsatisfactory regulations which overburden industry and the bureaucracy. This is part of the same problem. The legislature wants headlines, the executive gives the order. They may not have the resources needed to do it; it's all jumbled up. If the public were educated well enough to appreciate the need for extra money, we might pass the necessary legislation to succeed.

GRIESBACH   I came at the law/environment problem from a growing interest in, legal philosophy, in the foundations of law. I'm, of course, not alone in this. Over the last 15 years or so, academic lawyers have been obsessed with problems of theory -- a clear indication that something's wrong -- with school after school forming programs, only to be replaced by other schools. This has happened with law in economics, in deconstruction kinds of theories of law, to a critical feminist theory of law, and so on through a whole range of contemporary movements. In any case, it has struck me that the theme throughout much of this legal theorist soul searching is more of a generalist 20th century theme. There is an obsession with a kind skepticism and inability to have confidence that we know the good from the bad, the *is* from the *ought,* and so on. Essentially the problems in legal philosophy have been epistemological problems that have obsessed ethical theorists throughout the 20th century, if not longer.

Yet it seems that the epistemological problem is not the basic problem. The problem is more one of trying to figure out what we're talking about and what I call an ontological problem -- the problem of getting clear on what we humans are and how various social ills, troubles or objectives relate to that. It seems to me that over any period of time there are various conceptions of how we humans relate to what's around us, how we bear upon the surroundings, how the surroundings in effect constitute part of our vision of who and what we are.

In my paper I attempted to set out a rough sketch of three conceptions of the human context which I think bear upon environmental problems today. More, I think they in fact shape environmental problems in some respect and also shape our visions of how we can deal with these problems. I don't think it's useful here to try to summarize all three of the conceptions. Let me say just a few words about each.

The first conception I mention is that of the human as a self-defining individual, which I feel is fairly clearly an Enlightenment conception. It found expression in a great deal of 19th century American law. I'm not sure how far it extended beyond the United States. This conception, particularly in its vision of property as literally part of (an Aristotelian kind of property) the individual continues today. As those of you who are from Missouri know, there is a current debate over an initiative proposal on the November ballot to set up a system for planning for the streams of the state, providing for some sort of long term erosion control, some restrictions on the use of the streams and so on. [Ed., this initiative failed by a large margin at the polls.] In any case, the initiative has generated acerbic, even violent reaction among many landowners along the streams. Underneath the reaction is a Lockean conception of property rights, namely, that any interference with my land is somehow an interference with my person. At one point in the paper I sketch out the notion that any and every restriction on any property right is seen as threatening every property right. Even a modest attempt to set up a regulatory system finds that kind of a reaction under that kind of conception.

I refer to the second conception as a utilitarian conception. I think it's a prevalent, explicit or implicit, view of dealing with controversy. It's the view that social institutions are established essentially to maximize the satisfaction of our diverse and varying wants and so on. Under that conception we can see much of today's pollution control regulation generally as justified. The argument is justified as accommodating diverse interests, pressure groups and so on. One of the main problems, I think, is that it's ultimately self-refuting; it drives itself into increasing regulatory detail. I tried to sketch the extent to which I think cost-benefit analysis can be grounded in the same kind of approach.

Finally, I attempted to sketch in very skeletal form what I think may be a developing conception of the human context. I call it a constructural conception, under which we look at ourselves as not simply consuming nor self-developing individuals but rather as part of larger structures influencing and shaping those structures. The structures then shape our own modes and terms of existence. I point to, for example, the Endangered Species Act, the Wilderness Act, and so on as expressions of that kind of view of the human.

My overall point is that there's a need to look behind whatever kind of regulation or legal system we've got in force and attempt to identify what view of the human is being expressed therein. We're ultimately pushed to that kind of question in our attempt to justify any kind of regulation of environmental problems.

KANE When Bob Collier asked me to participate in this workshop on technology and the environment, the concerns he mentioned with respect to the environment varied with the work place. He and I are in genetic engineering, in biotechnology, for Monsanto, a chemical company. These two particular terms, *genetic engineering* and *chemical*, apparently have very definitive meaning for people. "Chemical" is synonymous in many people's minds with toxic chemicals -- "there are no good chemicals." Genetic engineering is a scientific process that is often seen as leading to very serious problems. I was in a meeting this morning with a fellow scientist who had gone to a Methodist church meeting where biotechnology was viewed as usurping the power of God, trying to fool with nature and doing things we shouldn't do. That's the context in which I was thinking about this particular write up. I tried to describe a little bit about the technology -- what it is, that it really isn't that new, that we have been practicing biotechnology for literally thousands of years.

Very early civilizations made wines by biotechnology processes. They used a biological organism to make something, and that's really all biotechnology is. What's the new thing? Why are people upset about it? We have a new technique. We have a technique we never had before and which has been given the title of genetic engineering. If we stop to think about it, we've been doing genetic engineering for a lot of years. On the farms, for instance, cows are now giving much more milk now than they ever gave before. Steers grow to a larger size. Plants give more corn and more wheat than ever before. We see different colored ornamental flowers. All of these advances are in the realm of genetics. They're sculpted by engineering plants, by crossing plants, by typical kinds of breeding. What's different is that we now have a specific means to take certain traits and transfer them. This technique allows a much more controlled kind of genetic engineering because, rather than mixing together genetic information from two different organisms, we're now taking a specific piece of genetic information and moving it from here to there.

In one sense I would suggest, as Dan Bennett did, that there's a lot more to understanding it than what we read in the newspapers. It's a process which seems to intimidate people and scientists certainly haven't done much to alleviate that problem. Having worked in academia and written grants for the National Science Foundation and other granting agencies, I can't ever remember once explaining to anybody why they were paying their tax dollars to a particular agency to pay me to do what I was doing. There's a good reason why people are concerned about scientists. In a way we have lost our pedestal, if you will. The process or the technique we now have is just that -- it's not a panacea. It's not going to solve every problem we have, be it medical or environmental. But it does present a unique opportunity. There have been some rather dramatic results.

In the pharmaceutical area, I mentioned in my paper a protein or a product called human growth hormone. Individuals suffering from dwarfism had to rely on doctors extracting human growth hormone from corpses; this then was administered to these patients. It's a very limited source, as you might imagine. It also is fraught with problems from viruses and all sort of contaminants. Now, by the technique of genetic engineering, we're able to make human growth hormone by a fermentation process much like that used by Anheuser-Busch to make Bud Dry (a beer). Human growth hormone is now rather easily available. In fact we have it in relatively unlimited amounts. This is a marvelous result of this particular technology.

Then let's look at the environment. What is the environment or what are the environmental issues? They're related to chemicals in the environment. They're related to toxic wastes. We hear Meryl Streep and others speaking about the dangers of all pesticides. I'm not trying to put these people down. These are obviously perceptions that they have and about which they feel very deeply. Nonetheless, they usually do not see these things in context. We need to understand how all this fits together. The genetic technologies may, at least in some cases, create some situations where our application of pesticides, for example, will be significantly reduced. That's a positive aspect of this particular technology. Perhaps the first application of this will be in toxic waste dumps or in waste site application. If one were able to use this technique to encourage an organism to degrade waste, the acceptance of the technology would be easier to get across to people.

But there are two sides to it. The public does not understand and appreciate biotechnology and is fearful of it. Yet what we're trying to do is to solve at least some problems arising from the use of chemicals in our society. In my essay I propose a couple of possibilities. In the business section of the newspaper a few days ago, there was a description of cotton which now is highly resistant to worms. The farmer no longer has to rely on pesticides. For us, that's a genetically engineered triumph.

Others see the connotation that suddenly there's something alive out there that was never there before. I think it actually goes beyond that. I think people aren't so much afraid of that as they are about where this technology will lead us. There's a major effort going on in sequencing the genome which will bring up all sorts of new moral/ethical questions related to the technological advances. But here, again, I am using this as an opportunity to show you how genetic technology may have an impact on the environment, at least as it relates to chemical/medical applications.

SHEAHEN   My paper had a lot of pretty pictures which I shall show on the viewgraph in living color. Several of the other papers seem to tie in well with some things I mentioned in my paper. I end my paper with five propositions which we might discuss later. They deal with the idea that the principal way humans interact with the environment, by and large, is in relationship to energy.

The first proposition essentially says that we have to consider energy if we're going to look for a solution to the environmental problem. The second -- I'm sure this is the most controversial statement on my list -- is that nuclear power is the way to solve the greenhouse problem. The third one has to do with underdeveloped countries. Most of what I'm going to say will focus our attention on this point: they need a lot of energy. The fourth point follows from that. If we don't do it for them and do it right, they'll do it themselves and do it badly. Finally, the fifth proposition is that education of the public is the most important task we scientists have, since so many good solutions to our energy and our environment problems are locked up because of public antipathy.

In the paper I emphasized predicting the future. It's important to recognize that getting it wrong cuts into the resources with which to do

the job. Every mistake in a place like America has serious negative consequences for underdeveloped countries. I want to make a big point of that.

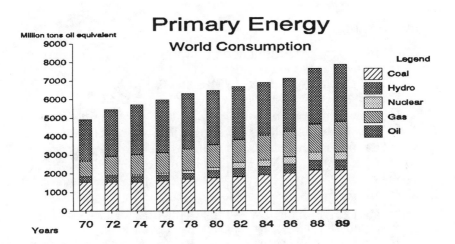

This figure shows what's going on in the world of energy. So in the time
from 1970 to 1989, world consumption of energy grew from 5,000 to
8,000 (million tons oil equivalent). This is about a 60 percent increase.
Oil is the dominant energy source; gas and coal are very important, too.
Hydropower is benign. I see nuclear as benign but others disagree with
me. Anyway, these latter two don't have CO_2 effects. They add up to a
pretty small amount. The rest of these -- coal, gas, and oil -- all generate
CO_2 and have an effect on the greenhouse situation. Here is the intimate
connection between energy and the greenhouse problem in the environment, namely, the generation of CO_2 from fossil fuel.

This next figure shows where all the oil came from. Some of this data is
excellent. This one is a good example. During the '70s when dependence
on OPEC was enormous, they could charge top prices. About 1979, their
capacity was huge. A couple of years later, these other sources -- the
United States, the rest of the world, the Soviet Union -- increased and the
demand dropped; there was a glut. This kind of connection between
energy sources and economics drives an awful lot that goes on. Once
there was a glut of OPEC oil and the price began to drop, American and

Japanese money became available to fund loans to the Third World. The difference between what we can do economically in, say, foreign aid or helping undeveloped countries is often closely related to questions pertaining to energy supply.

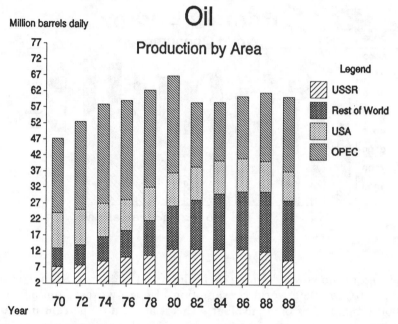

There's talk about special energy sources (solar, tidal, geothermal); they make good copy in the Sunday supplement. In 1988 in the U. S., fossil

U.S.A. Energy Sources
1988 Production

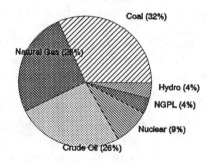

fuels dominated: a little nuclear, a hydropower and NGPL (natural gas and petroleum liquids). You can't find wind, geothermal or solar.

Synthetic fuels are interesting. We have more coal than we know what to do with -- a 600 year supply. The idea behind synthetic fuels is to take some of this coal and convert it to the equivalent of oil. Our hope for the very, very long term in this country relies heavily on coal.

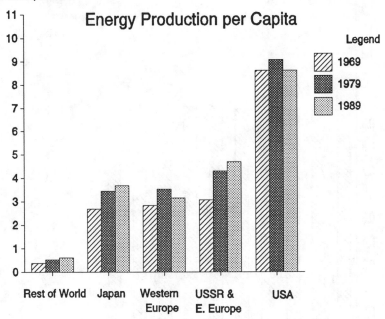

The next figure is extremely important. It draws our attention to the difference between what we have in the U. S. and what the underdeveloped countries are facing. We have a certain amount of production, pretty decent, but our consumption is very high. How do we get from production to consumption? We import. We export a little bit of coal. Happily the Japanese don't consume as much as we do, but the idea is the same. A poor country, Bangladesh or some other place, is very low. The most egregiously wasteful country is the United States of America. The entire effort of energy conservation over the last 15 years has been somewhat effective in that we've lowered the 1989 column down to the vicinity of 1985.

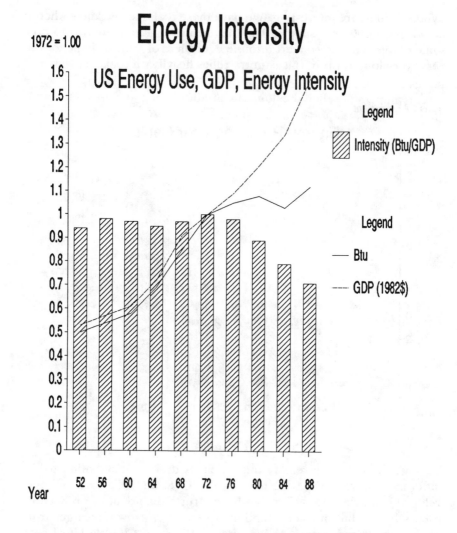

One thing you can do in energy use -- a company like Monsanto is very much aware of this and I think we want to focus on this as we look at numbers of the overall production versus the amount of per person -- is improve the parameter called energy intensity which is the energy per unit output. Long ago the energy (the solid line) and the gross national product or domestic product (the dotted line) tracked along together. It was widely believed as recently as the '70s that these two were in lock step. It was thought that, as the gross domestic product went up (which more or less increases with the population and with consumption) energy use would track right along with it. In all those years, the energy per unit product (these bars here) was pretty much constant. It normalized at 1.0. We discovered energy conservation in 1972 when the price of energy went up. When the Arabs hit us with a price rise, we discovered energy conservation. Since then the domestic product kept going up but the total energy consumed leveled off, which means that the energy per unit product started down. Lately (1986, '87 and '88 -- remember oil hit about $9 a barrel in 1986) we haven't done much better because the price dropped. But when the price goes back up, as it has been doing in the last couple of months, and indeed in the last couple of years, we should see a continuation of this downward trend in the future.

We're essentially saying to the underdeveloped countries, who are in real squalor and have nowhere near our energy use: "when you work on it, we'll meet you someplace in the future." If we leave it to them, they're going to be very high in energy intensity. But if we work together with them, we can meet at a point out here with more products and less energy. That, I think, is a national goal and it's this kind of thinking that underlies the points three and four that I mentioned at the beginning.

I want to show this next figure because herein is what I have to say about the incredible expense associated with a mistake. Everyone of these are reasonably competent forecasts, one of which is from a government agency. They're all pretty capable. Look at the numbers made in about 1985 and even in 1990. There's a significant spread. But if you can build nuclear reactors or coal burning plants in 1990, and you've got to make the decision if they will be effective in 2010, look at the spread. This is practically a factor of 2, (2.2 up to about 4.2, depending on who you believe). If you run a utility or something, you've got to make your decision, you've got to go to the bank, you've got to get a billion dollars to erect a coal burning plant or anything. If it turns out that you have undercapacity, you're in big trouble with your population. They're going to be mad at you for brownouts and so forth. If you have overcapacity, your stockholders don't make money. During the 1970s and before, the tendency in America was to make big expenditures to build more nukes than we needed. This cut into the amount of banking money available for infrastructure in the rest of our country. And it's a real shame that we can't do it right. That's the way this part of the prediction goes.

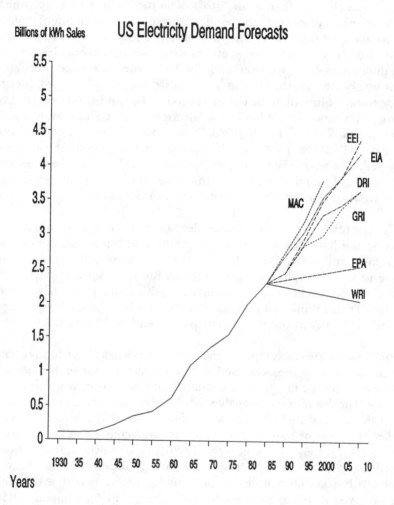

MAC -- Siegel and Sillen
EEI -- Edison Electric Institute
EIA -- Energy Information Administration
DRI -- Data Resources, Inc.
GRI -- Gas Research Institute
EPA -- Environmental Protection Agency
WRI -- World Resources Institute

Let's return to the graph I showed you before [page 145]. This shows energy use in 1969, 1979 and 1989 for the rest of the world, for Japan, Western Europe, the Soviet and Eastern Europe and the United States Look where we are. We can agree that the Japanese have a decent standard of living. Why we can't get from where we are down to the Western Europe level or the Japanese level. What are we doing wrong? Is it the gas guzzlers getting 10 miles per gallon or something else? They have a better standard of living in Japan and Western Europe than the Eastern Europeans do. These latter are higher in per capita consumption because they're wasteful and inefficient in their use of energy.

The future lies in bringing the "rest of the world" up to a decent standard of living. For us as a nation to stay way up in our energy consumption per capita demands our continuing to break the backs of the "rest of the world." That's a truly rotten way to run the world. As Christians and as scientists, as Americans, too, we should help these people and raise their standard of living. Also, a per capita chart conceals the fact that there are a lot more of them than there are of us. There are 3 billion people in the "rest of the world," maybe 4 billion; there are 250 million in the U. S. Future American leadership, certainly assisted strongly by the Japanese and Western Europeans, demands the creation of an equitable world in which the "rest of the world" is much better off.

If we do that by burning fossil fuels, the increase in CO_2 is going to be truly gigantic. If we don't increase the CO_2 that much, then these people will remain as they are, unless we find another energy source that doesn't produce all that CO_2. That brings me to point number two, namely, that nuclear power is the best way to solve the greenhouse effect. Throughout all of this we have a public both in this country and in the world -- a media, a TV system and so forth -- which is, practically speaking, scientifically illiterate. This brings me to my fifth point. We have a big job ahead of us in informing the public. The main thing I have to say is that the dichotomy between what America has and what the underdeveloped countries want and are entitled to is very great. That needs to be addressed in any global solution to the environmental problem.

LEGUEY-FEILLEUX   I'm glad that we switched the order of these presentations because I'm very comfortable with what I've just heard. It undergirds the point I want to make. My concern is about the global environment. I want to stress that it is not my intention to neglect the other levels of environment degradation. In the presentations, we had a spread from the very local to the global. Some forms of urban blight are at the neighborhood level. They are created at the neighborhood level and can be solved there. My concern is that the environment at the global level is one more area in which the nations of the world are enormously interdependent. Unfortunately, our political structures and ethos are still very much in the past. We belong to the 19th century and earlier. We are state-centered. We are parochial in the worst sense of the word. We live in a global society of which, at times, we are totally ignorant. We don't care about what goes on beyond our boundaries. Even when we make pretenses of working with others, our main concern is to pull the blanket to ourselves and push them out of the bed. The trouble is that we're all chained together -- if they fall, we go with them. That's the problem I'm concerned with, not detracting from the other dimensions of the problem.

We are lucky that we don't have to reinvent the wheel. Pushed by need, our leaders set up a world mechanism. Remember that the U.N. was created by people like Churchill and Stalin and Roosevelt; we can't call them idealists. So often in so many of their policies, they were cynics. They were power politics experts. They said: "Well, if we're going to avoid in the future what we have just gone through, we've got to have new structures." That's how they came to the idea of creating the U.N. The U. S. had refused to join the League of Nations and the Soviet Union was expelled from the League. So they were not exactly enamored by international structures. Yet they said: "Somehow, let's do something that will do the job that we tried to do before."

So we had a beginning. We started very late. The first international structure to deal with our interconnectedness dated back to the last third of the 19th century. A few hesitant steps were taken then. There was a bit of experimentation between World Wars I and II. A good example is the ILO (International Labor Organization) which was pushed by the specter of a labor revolution. There were no high ideals involved, that somehow we owed it to the workers to give them a better life. Rather it was feared that they would get up in the middle of the night and slit our throats. So we thought that we had better do something about it. We invented a structure where the workers would have the novel role of participating in equality with management and with government representatives. So there was some experimentation. Nonetheless, it was after World War II that we really turned to international institutions.

We were really concerned about the environment before it became a global concern and there were ad hoc institutions created to cope with problems that were staring us in the face at the end of World War II. Eventually we realized that we had environmental problems. We didn't call them that then, but we started working on them. Take, for example, the World Health Organization. If we wanted to foster public health globally, we had to be concerned about environmental health. UNICEF's concern for the well being of children and their mothers and about infant mortality led to deep concern about the quality of surface water. These people loved it. It tasted better than what they could find in the ground. But it was killing them. UNICEF launched a global clean water program of ten years in cooperation with WHO (the World Health Organization). They were working on one dimension of the environment.

The population explosion is a very touchy issue. The First World tried to impose its solution on Third World peoples and they imaged that we had all kinds of hidden agendas. We probably didn't, but they thought that we did. So the UNPA brings everybody together and says: "Why don't you people start talking to each other. Maybe you'll learn." Now people say: "We must have more international world population conferences because there's something for us to learn from this exchange."

That has given us a base to build on. Let's realize that this was all piecemeal, each agency doing its own program. We had to wait until 1968 to get a global view of the global environment when Sweden, plagued by its own environmental problems, started getting after the other nations diplomatically. "Let's do something together globally at the U.N." That's where the lobbying was done. That led to Stockholm in 1972. Sweden was the host, because it had been pushing so hard to have this gathering start talking about a global environmental strategy. That conference was a great catalyst. It invited those other organizations to redouble their efforts. It invited organizations that had not been so involved to start considering the environment. For example, the World Bank was ready to increase the amount of money it was investing on environmental dimensions of the projects already being financed. Before Stockholm it hadn't been that concerned about the environment. Now it was prepared to put money into environmental projects.

The Conference was a great "legitimizer." The environment had been the preserve of a very small group of people who were not always thought to have it all together. All of a sudden the world got together and agreed to put together a plan of action. That's quite an achievement, especially when we involve the Third World. The Third World used to view the environment as the rich nations' sin, saying: "When we get to your level of development, we can worry about the things you're worrying about." But why wait, since you (the Third World) can achieve more development if somehow you preserve your environment? But how do we teach that?

The world environment conference in Stockholm started this progressive chain. They created an agency that has remained small, the U.N. Environmental Program. It, too, has been a catalyst. Its budget is only $5 million a year for administration; they raise $30 million on their own. With the collecting of reliable data which could be communicated, we could use standard units of measurements. There has been great difficulty comparing statistics across international boundaries. It has done quite a bit. The bottom line of my paper is that it's not enough. We still quibble over U.N. costs. The U.N. costs us (the globe, not just the U. S.) just over $4 billion a year. Our U. S. budget is way over the trillion dollar mark for just one country, and the world has to be served. The U.N. is spending $4 billion for everything, only a tiny fraction of that for the environment.

I hope we come to our senses. Indeed, we have to be concerned about the Third World. To do that we have to change our ethos. So long as we remain state-centered, so long as we are governed solely by our national interests, we are pulling the blanket to ourselves. It's time for us to realize that a better global environment will be a better environment for us as well. True self-interest is one that fits into the global common good.

GENERAL DISCUSSION

SESSION II

KARLSTROM   I teach physical geography and I have some comments and questions. I'd like to start with a few comments on climate. I found very little to quarrel with in Ben Abell's presentation. He did a very good job of presenting a complex subject. If I could, I'd like to add a little bit. I'm going to be a little braver than Ben and mention some numbers so that we can perhaps assess the relative magnitude of natural versus human-induced climate change. I'm going to draw on the vertical axis millions of years, 3 million years ago; and here let's start with the average climate of the world today which is about 15°C.

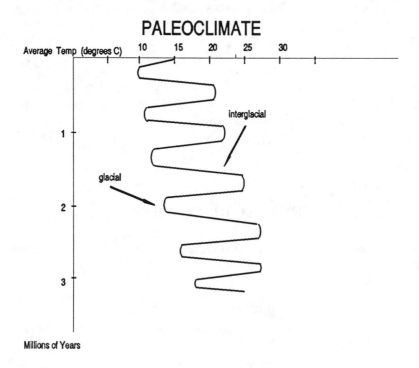

This little diagram is based on a lot of paleo-climatic work. The temperature limits I'm using are from some European work on distribution of fossils and plants. We know that reindeer lived in Spain during the last ice age. We know that warm species (plants) were living at the very top of Eurasia during some of the past interglacials. By mapping these plant migrations, we get a sense of what the climate probably was during the last 3 million years. The average climate at the end of the tertiary, the beginning of the ice age about 2 or 3 million years ago, was about 7 or

8 degrees Centigrade warmer than present. The average world temperature was warmer; the oceans were higher. There was much less ice in the ice cap. Since then we have experienced a series of maybe 20 to 30 glaciations in the last 2 to 3 million years.

It appears that there has been a general cooling trend; each successive ice age has been getting colder and each interglacial has been a bit cooler than the previous one. During the last interglacial it was a bit warmer than it is now; we have fossils of giant land tortoises which lived in Illinois 125,00 years ago. These now live in the Galapagos Islands. These land tortoises cannot survive freezing temperatures. Their having been in Illinois shows a dramatic change in climate.

It seems that during the earlier interglacials the climate would go back to that of the tertiary. We're now in an interglacial that has lasted about 10,000 years. Interglacials would be on the order of 3 to 7°C mean annual temperature cooler than in the Tertiary and glacials would be colder. There's a difference between a continental locality or one next to an ocean. I'll consider a continental locality, since I did my work in Montana. Montana might have been on the order of at least 10°C colder. Some of the work suggests that we have a minimum temperature difference in Europe and in Montana of 17°C between glacial and interglacial. That's a whopping temperature change on the earth. That's natural change, not human-induced; this occurred for the most part before humans could change much.

In 1880 we had 280 parts per million (ppm) of CO_2. In 1990 we have about 354 ppm. We have increased the temperature by about 0.6 degrees Centigrade, a little over a half of a degree. Whether this warming is due to our burning of fossil fuels is unknown. It may be a little bit, but even so, it's a very small amount compared to the natural range of climate.

I think that we have to stop burning these fossil fuels because they are nonrenewable. It's an unrepeatable gift, if you will, and we're going to be out of them in 100 years if we simply burn them. Yet I don't believe that we are justified in being so pessimistic about what humans are going to do to the climate since nature seems to have much greater impact. I'll throw that out as a place from which we might begin a discussion.

SPITTLER   I'd like to comment on that, from a perspective of a chemist. We need to be much more careful in the use of our hydrocarbon resources, because, they are tremendous resources for chemistry. To burn them as fuels is an incredible waste of something that's a valuable raw material for probably thousands of years instead of a hundred. I've had the same suspicions about the CO_2 cycle that you suggest, although I didn't know the long range effects.

Accurate weather measurement dates from not much later than the 1800s. One of the earliest people to keep any kind of records on a day-to-day basis was John Oliver. He kept a weather record for about 60 years of his life. At that time, there were just individuals keeping records. We don't have any kind of accurate weather records for the global trends in most places for more than 50 or 60 years. We don't know the details of the weather over the last 100 years; temperature changes in most countries of the world are uncertain because we don't have access to that data. Making highly confident statements on the basis of the information we have is, at best, very unscientific.

BENNETT   I have a question for Dr. Leguey-Feilleux. The U.N.'s International Panel on Climate Change seems to be pretty unequivocal in its analysis on global warming and it's hard to get direct commitments from the major industrial countries which is so important in addressing ozone and greenhouse gases. Do you think that they have enough information to make such a strong stand?

LEGUEY-FEILLEUX   Well, it seems that, even though some of the transformations may be due to natural causes, the release of large quantities of CO_2 must have some kind of an impact. These are only predictions, and committees have been proven wrong in the past. I think that in light of our increased consumption of energy and the release of CO_2 due to the material we use, there will be a warming effect. I think we ought to remain concerned about the impact it has, even if the predictions are only approximations at the present time.

KARLSTROM   That's a good point. I want to stress that on the one hand natural climate change is much greater than anything we have accomplished or probably will be able to accomplish in the next 100 years. On the other hand, the global warming debate is a very positive thing. If it scares people into changing their behavior, it's a positive thing. If indeed we do warm up the climate 1.5 to 4.5°C on average by the year 2050 -- one of the projections -- we will have changed the climate enough to impact greatly our own culture and civilization. This is something that

I don't want to minimize. In fact, they predict that the impact would be greatest at the high latitudes, maybe on the order of 10 to 12°C warming, which would dramatically reduce the ice cover on the poles and raise the sea level. So it's something we need to take seriously. But as Ben points out, there are so many factors other than simple CO_2 content of the atmosphere which could change climate that these projections are a little bit shaky.

GRIESBACH   In your paper, Ben (Abell), you mentioned that water vapor or variations in water vapor content might dwarf the influence of CO_2 on the greenhouse effect. Can differences in water vapor be attributed to human activity or is it uninfluenced by human activity, an aspect of long-term climate changes like glaciation?

ABELL   Yes, to both. We have tremendous variation. The water vapor content is variable from place to place; it's much higher, for example, in the equatorial climates, minuscule in the polar climates. It varies with the season and probably varies from decade to decade. Much of the climate appears to be episodic. If we look at the climate in terms of only one of the episodes, we're liable to come up with erroneous conclusions. This is true whether it be the cooling in the 1970s or the warming of the first half of this century or what appears to be some renewed warming in the last ten years. Still, the trend appears to be toward warming. But if we use just one or other episode to make predictions we will probably be way off base. We need a longer data base.

We're mapping water vapor very well now via satellite and infrared technology. We've been doing it very well for about 10 years. We were doing it experimentally from the second or third Tyros experimental satellite back in the '60s. We've been doing it on a regular basis since we've had the GO satellites available. But 10 years doesn't make a climatology. We're going to have to wait 50, 60, or 100 years; then we'll start to see a climatology as far as water vapor is concerned.

Yes, the water vapor content is variable, but human activity could have an effect on it. Anything that would produce a warming could promote more evaporation. With more evaporation, we get more water vapor in the atmosphere. More water vapor in the atmosphere will accelerate the greenhouse effect. Some people have argued toward a run-away greenhouse effect because there's a lot more water vapor in terms of rain in the atmosphere than there is CO_2 -- or will ever be, for that matter.

But, as I pointed out, if we get an increase in water vapor, there's a strong potential for an increase in cloudiness and in precipitation. If we increase the cloudiness, we increase the albedo and produce a cooling to counteract the warming. We just don't know where we are.

BERTRAM   When you say climate changes are episodic, does this relativize in your judgment what Eric Karlstrom was saying about the regularities of these patterns over the millennia?

ABELL   No. His point goes beyond the episodes. My point is that I don't know what the climate is doing now or in the last 10 years. I'm seeing ups and downs. Let's go to the year 2000, 2010 or 2020. Then we can look back and get a better perspective. I question even the present data base. I'm a little surprised at the magnitude of your (Karlstrom) temperature variation, but that is, I guess, very solid data. That has nothing to do with these episodes. My problem deals with weather and weather prediction and with climate which is sort of an aggregate of weather. Everybody likes to say things are governed by the jet stream. It's not quite that simple. We have wave patterns in both hemispheres. By the way, most of our data is coming from the Northern hemisphere. In the Southern hemisphere there's a lot of water and we're talking about developing countries down there; their data bases are not so good. We're dealing with this from the point of view of the Northern hemisphere. We had extraordinary cold in the Ohio Valley in the late 1970s while at the same time it was extraordinarily warm in the Rockies. These patterns shift and we can't even tell why that is occurring. We can give only a part of it.

If we get cool temperatures in the upper atmosphere, we'll get low pressures there. There are no ifs, ands, or buts. So we are going to superimpose waves on a hemisphere. We'll get a deep cold pool over the Hudson Bay every winter. We'll get it over Greenland and Siberia. So if we get low pressure in that area and if we get a circulation around the low, we automatically impose a three wave pattern on the Northern hemisphere which sometimes can increase to as much as five, six, or seven waves due to energy inputs that are not completely understood.

If that happens, we'll see tremendous variation from one winter to the next winter to the following winter, not only in a given location perhaps but maybe over the whole hemisphere. We'll see more of a meridional flow pattern where some people are going to be extraordinarily warm and other people are going to be extraordinarily cold. We're dealing with the available data, data with which we have to be cautious because of the station selection. We're trying to find a trend and I'm not sure I can come up with one as far as the 1980s are concerned. The data base is imperfect at this time; that's certainly true of data on water vapor. We can map water vapor very well for the last 10 years, but that's all. We need more and better data; we need additional sophisticated (and expensive) ways to handle the data once we get it.

COLLIER   There's an unstated premise running through this debate: "it is desirable for change not to occur in the environment or in the animal populations. As they exist now, they should be frozen in time and we should forever maintain the exact population and environment that we have now. If the human population did not exist, these environmental problems would not arise." We know from the fossil records that the world has gone through massive changes, so I do not accept the position that change should not occur. Change is going to occur whether we want it to or not. There is nothing inherently evil about the environment changing. It has been changing for as long as we know.

The debate about environmental change in the end revolves around population growth and whether there's inherently something evil about population growth. This issue in the debate hasn't been stated yet, but it gets to the heart of the whole thing. Many people say that the growing world population adversely affects the environment and is inherently evil; therefore, we should do something about it. I want to raise that point.

LEGUEY-FEILLEUX   Some countries find it difficult to invest in more data. When we have a crisis on our hands it's easier to take drastic measures and start spending, but gathering data is not spectacular. That must be done year in, year out. This is a place where we could invest more of our funds in UNEP, the U.N. Environmental Programs. They are funded, but they're always short of money. UNEP has been a catalyst to bring about more data gathering by inviting states to create their own stations and to join in a data network. We could put more of our resources into that global institution to generate accurate data. It has been one of its main missions to make sure that the data we have is more accurate using the same standards of measurement and exchanging data.

SPITTLER   Ben's response suggested something with both a positive and a negative aspect, namely, how these projections are made.

They are made using a computer simulated structure of the environment, and that simulation is as effective as the data put into it. We're in the infancy of using these computers. The capabilities of computers are sometimes rather highly overrated in terms of what you get out of it and what you put in. We assume that we're putting in enough data for the computer to come up with a simulated range of values of changes of temperature or prices of oil or whatever it is.

Positively, now when some of these things are beginning to affect the global environment in terms of human environment, we have a technique with the potential for dealing with it. Negatively, we can put far too much credibility into a program that depends on the quality, the level and the amount of information as well as the assumptions behind the simulated program. Sometimes these last are extremely oversimplified. To do quantum mechanical calculations of even simple molecules, we need to make drastic simplifications in the mathematical formulae. When we're dealing with long-range weather patterns, we're dealing with mathematical formulae which are, absolutely horrendous, especially on a global basis. We're dealing with incredible complexity. We shouldn't think that, because we can make a simulated program for this, we can trust the projections. We have to be cautious in extrapolating what we've got.

BERTRAM   Do I detect at least a kind of symmetry in all the remarks? On the one hand, the warming effect, at least on what information we have, is detectable enough to warrant our intervention. On the other hand, the data is so massive and so unknowable, at least with precision, that we have to be very cautious about the predictions we make. Am I oversimplifying?

ABELL   We oversimplify almost everything we do in a meeting of this kind. Yes, we're possibly seeing warming. We certainly have experienced it this century, although more in the first half of the century. A major part of that could have been caused naturally. I think that doing anything drastic at this time, as I said last night, would be very premature. I liked Father Spittler's comments on computer modeling. How good, how representative is the data? I agree with what he said about assumptions as well. Not only is the output very dependent on the data that goes in, it's dependent on the program as well. We're dealing with imperfect things here.

I've worked on tornadoes for the last 10 to 15 years, trying to come up with a tornado model in terms of causality. Some computer models have come out using the most sophisticated data techniques that we have now, where we're combining ground surveillance with various types of radar surveillance and with satellite surveillance. The model that many people accept right now seems nonsensical. It looks to me almost as if they've developed circulation from the ground up rather than from the clouds down, even though at times visually it appears to start at the surface first. So we're hurting there. We can do wonderful things with computers. We can handle a lot of data in a very short period of time. But I've run into problems like this time and again, and I've become very careful in taking modeling and starting to run somewhere with it.

KARLSTROM   I heard Stephen J. Gould, the chairman of geology and biology, I think, at Harvard, say, in a speech at the University of Colorado, that ascertaining the long-term and short-term patterns of climatic change is perhaps the most difficult task facing science. There are so many variables! The problem with the computer models is that they try to limit the number of variables. As Father Spittler said, we may or may not come up with an accurate simulation.

I would like to pick up on Bob Collier's point about population. We can't talk about the environment without addressing this. It's extremely critical. At 5 billion, are we approaching the carrying capacity of the earth for the human species? Some ecologists (like Barry Commoner) believe that we are and that population is *the* most important problem. Whether it is or not depends on our use of energy and on our impact on the globe. How we live depends on what we think and what we believe. This is why we're all here -- trying to put together what we believe with how we live. I think that the only possible solution is to modify our behavior to conform to a set of accurate beliefs about our relationship with the environment. Population is something we must look at.

SCHNEIDER   To return to the global warming for a moment, though it would be applicable to some of the other environmental questions as well, it seems that there are two drastic extremes to be avoided. One would be doomsday scenarios and running off in a direction before we have adequate data. The other extreme, to my mind, is to fall back into saying, well, because we see dramatic change on the earth over geological epochs that change is, therefore, automatically acceptable. We're asking a question about possible effects over our lifetime and that of foreseeable generations. The ultimate question seems to be social rather than technical and scientific. In the absence of good data that can be obtained in the very short run, how do we act in a responsible fashion in the light of the data we possess? That touches on the kinds of things that Eric has said about sustainability. It also includes what Jean Robert touched on in his paper and in his remarks about a global society or elements of a global society acting responsibly in the light of the present situation.

EVERETT   I'd like to comment on change being good or bad. Change is life. We should address the rate of change. If change occurs extremely rapidly, if the natural change that is with us anyway is so accelerated by human activity that change comes faster than we can deal with it on the social or economic level, then change is bad. How we can keep the rate of change at an acceptable level is the question. That ties in with population. I lived in Brazil where the rate of population increase was extremely rapid. They now have too few schools. The government set a limit of no more than 50 children in a primary class. In practice you will find maybe 52 or so, putting a few extra in just to have them there. Older children are not going to school at all. There are no teachers or buildings. They can't handle it. We can't have a world without change, but we can affect the rate of change. That needs to be looked at.

LEGUEY-FEILLEUX   In response to Sister Maxyne, indeed there are all kinds of things we need to do differently. Last night Tom Sheahen showed a figure on our manner of using energy. It's scandalous that, with the available technology, we can still be so wasteful because it's the line of least resistance. I agree with regard to the population: what is the level of sustainability? But we ought not to address ourselves solely to the Third World which has high population growth rates. With a lesser level of population growth, we consume at an enormous rate. We have to look at both sides. It is necessary to curb that very high rate of population growth when people suffer as a result. At times this occurs because of a fouled up conception of the family, like regarding it as an old age insurance. What a lousy way of looking at one's family.

EVERETT   Some don't plan it at all; it just happens.

LEGUEY-FEILLEUX   That's right, but some are very conscious of it and don't mind talking about it. This is not the perfect way of looking at one's children. Conversely, we seem to be so smugly satisfied ith the way we do things that we see no need to change. Of course, we should be the easiest ones for us to change. We ought to concentrate on what we do incorrectly and work harder at changing it.

CROSS   In their book on games theory, Von Neumann and Morgenstern point out that the study of the weather is the second most difficult problem that science faces. As a layman with respect to meteorology and climatology, my big confusion is in two areas. I hear about evidence that particulate matter in the atmosphere has a big effect on climate; and that slash and burn farming, related population growth, and so on, are big variables. I also read that there is evidence that volcanic eruptions have had profound effects. This is sort of a natural experimental approach to understanding climate as opposed to getting a firm scientific base. I'd like to learn more about that kind of data and how much of a worry such things should be.

I'm also confused by the use of numbers without errors of estimates. I try to understand statistics and teach it to psychology students. That's the most difficult area, trying to say what the variability is around these estimates that we make. It seems to me if we heard more of these kinds of figures, we'd be in a better position to know whether there are dangers or whether there is no basis to think there are. That's some of my frustration. If the errors of estimate are really small, then we need more data about possible things that can be done and about the errors of estimates on those guesses, too.

COLLIER   I want to separate the social issues which Armgard very well pointed out with the population growth versus the actual "carrying capacity" of the world. I have seen estimates of a carrying capacity of 36 billion, assuming that people are adequately fed and housed. The separate issue of too rapid a population growth in one area or under one political system and their inability to provide for those people is a separate human issue. The issue I was referring to is whether or not there is any inherent reason to try to freeze the world's environment or animal population.

EVERETT   We can't do that anyway.

COLLIER   I believe that's correct; we can't accomplish that. Nevertheless, much of the debate around the environment and these issues has as its basic starting point that this shouldn't change.

EVERETT   That's right. I would like, therefore, to interject that absolute numbers are not the real issue here. It is how rapidly we are increasing the numbers, namely, the rate of change. And if the rate of change is exponential, then the change of the rate of change is also exponential. We are possibly getting into a run-away situation. If the equations that we are using are not linear, there are possibilities that very slight changes in rate produce completely different equilibrium situations. I get that out of chaos theory. The climatologists have gotten into the study of chaos. A small increase in one direction can cause huge effects in a relatively short time. That could be humanly unmanageable long before we reach 36 billion.

BERTRAM   Armgard, if I understand your point, the rate of change can be bad only if it proves to be humanly unmanageable and, therefore, intensifies, let's say, the consequences in human misery.

BERTRAM   Is there something like an acceptable rate of human misery?

EVERETT   That depends on one's point of view of the human being. As many totalitarian philosophies would have it, a human being is like a cog in a machine -- expendable. They can be put into labor camps or killed wholesale. There's no level that's unacceptable. If we regard the human being as the image of God, that's something else. We can start to philosophize from there and try to determine an acceptable level of misery. I wouldn't like to pronounce on that.

BLASCHKE   As an amateur in the business of the environ-ment, I confess that a sense of almost overwhelming complexity defeats me. I am uncertain about what we can do besides recycling aluminum cans, not using styrofoam cups and that sort of thing. The complexity is not only on global basis but on the microscopic basis. For example, the world of antibiotics and microbiology, (germs, if you will) is changed forever; it will never be the same again because we have better antibiot-ics. We've created a whole new spectrum of microorganisms that are different from what they were originally when they were discovered. An example is the hospital-based staphylococcus infection which is in many ways different than it was in 1890 and will be different in the future. With each new antibiotic discovery, the organisms will change.

We also have new kinds of infectious diseases -- AIDS, hepatitis C, and an ever growing list in which the microorganism, the microscopic world, is changing and we can't stop it. There's no practical way of stopping it. I find myself in agreement with Dr. Kane. If you've already read his paper, re-read it. It's one of the best discussions of the technology of interpreting DNA and the whole field of molecular biology that's dependent on the technique of restriction enzymes that he mentioned and in which we really find out how the human body and almost everything works in terms of biology. We're committed to a world of everlasting change.

How we humans can control it is where I feel a sense of defeat. It's as if we're caught up in a river of change, and for better or for worse, we have to make the best of it and keep going. I'm beginning to feel, after hearing Dr. Sheahen's talk last night, that finally we must come back to theology or the spiritual dimension as the only way we can change things.

McLEOD   A question came to mind as people were talking about the environment and the rate of change. That discussion brought to mind Carl Sagan's famous articles a few years ago about the "nuclear winter" if there was an all-out nuclear war. Sagan seemed to indicate that the temperatures would drop significantly causing havoc with agriculture and other things. Then I heard later that some scientists made studies and said it may not be a "nuclear winter" but something like a "nuclear autumn," which would still devastate agriculture. Is "nuclear winter" a sensational statement which blows these effects out of proportion or is there some sort of basis for this kind of climatic change due to particles being put into the air?

BENNETT   There's a lot of geological evidence that such things as volcanic eruptions have an effect on it.

KARLSTROM   Exactly. Krakatoa erupted in 1883 and sent a plume of volcanic ash into the stratosphere which stayed there for five years and cooled the climate. The leading theory of the extinction of the dinosaurs is that a planetesimal or a large asteroid or some body smacked into the earth and knocked up so much particulate matter that it blocked the sun's rays, thereby killing plants, the food of the dinosaurs, and precipitating their demise. There is, I think, a firm scientific basis for at least some of the projections of nuclear winter. More than that, the World Health Organization estimates that a couple of billion people would die, maybe three. So it seems to me that when we're talking about that level of human destruction as well as ecosystem destruction, we're almost quibbling when we ask, is it really as bad as they suggest. It would be the worst catastrophe in maybe 65 million years to the planet.

The scary thing is that people without a scientific background are toying with these ideas. There was a Harvard economist, I understand, who, after listening to these environmental problems, said "I wonder if global warming is such a big problem, if we could have just a limited nuclear war to offset the effect." I'm frightened by the assumption that we should somehow engineer and control these things. We are not in control. We like to think we are in control technologically and scientifically, but we're not. We're still one species on the earth, and the earth is still much more powerful and God's much more powerful than that. This is a dramatic example of hubris and arrogance on our part to think that we can engineer a world to our liking. We're seeing the failure of that arrogance, which I think is what the environmental crisis is teaching us.

COLLIER   The world amphibian population is dropping dramatically worldwide. We have no explanation for it. We can't tie it to any pollutant and, since it's in so many different locales, we can't tie it to a specific environmental aspect like temperature or exposure to "toxic elements" in the environment. It is a totally unexplained change and may be related to the population itself, or to some dissonance in the population or possibly to new forms of degradation that we don't understand. There is a study going on here in St. Louis using specific ponds. They've seen the same thing, that the amphibian population has dropped dramatically in the last few years. The local explanation is that flooding has brought in new predators. Well, we can explain that possibly locally but certainly can't explain it worldwide. Some of these things do happen without any apparent cause that we can tie to a human-induced thing.

SPITTLER   The Chernobyl experience in many ways could be a blessing in disguise. I don't know what impact -- I would like to ask Jean Robert about that -- it has had in the Soviet government's policies. This was *not* a nuclear explosion; it was a hydrogen explosion. Hydrogen is generated from nuclear reactions and the reactor exploded and the nuclear material scattered about the town. Yet look at the tremendous impact it had all over northern Europe even over Greenland!

My view is that one has to look at this from the perspective of divine providence. Maybe this accident happened through human error. Maybe it happened at a time when we needed something like it, in order to recognize what would happen if a really wide range of nuclear explosions took place. I agree with those who have talked before that there's no question that if a wide range of nuclear explosions took place -- probably some of the estimates are exaggerated -- it certainly would be a terrible experience and would take a long time to overcome. Maybe the accident is something that puts us in touch with what could happen if we had a serious nuclear explosion.

WELLMAN   Father, what do you mean by wide ranging?

SPITTLER   I mean if we had, say, a nuclear exchange.

WELLMAN   No. I mean about Chernobyl. When you say it had a wide ranging effect over northern Europe and Greenland.

SPITTLER   Well, all over northern Europe and Greenland, the effects of it have been picked up.

WELLMAN   There were traces of radioactivity. I don't agree that it was a wide ranging effect. It was widely detected, but that's because it's easy to detect infinitesimally small amounts of radioactivity.

SPITTLER   In the Ukraine, the effects were serious in terms of crops and things like that, at least as I understand it.

KARLSTROM   Are you trying to minimize the damage of Chernobyl? It seems that there was a massive caribou die-off in Lapland and maybe an estimated 6 to 12 thousand people are going to die. . .

WELLMAN   There was a massive kill-off because they killed the caribou. They didn't die naturally. I am just trying to clarify what he meant by massive. It certainly was detectable worldwide.

KARLSTROM   They killed the caribou because they were hot.

WELLMAN $Cesium^{137}$ contaminates the muscles from the fallout. Before the caribou vacuumed up all the $cesium^{137}$ and contaminated their muscles, they killed the caribou so it was still edible meat.

KARLSTROM   I don't believe they ate it, did they?

WELLMAN   To my knowledge, some of it they ate. I've been
to a number of presentations about the effects of Chernobyl. My field is
nuclear medicine, so I have some understanding of it. My original work
immediately out of medical school was with the Public Health Service
looking at fallout and how it gets into human beings. I'm not trying to
minimize the effect of a nuclear explosion. I appreciate very much Father
Spittler's emphasizing that it wasn't a nuclear explosion. It was a
conventional explosion or explosions, if you will. Even my fellow
physicians come down to the department and they kind of back into it, as
if we're doing some kind of hocus-pocus or something.

I only want people to be reasonable and scientific when they start talking
about radioactivity. It's an understandable physical entity. It's nothing
spiritual or demonic. People, even some scientists, tend to get fearful
when they start to talk about radioactivity and nuclear events. I'm sure
that a lot of the information that's coming out of Russia about Chernobyl,
stories about cows with two noses and 10 horns or things like that, is an
exaggeration and a misconception. It will be very difficult to keep the
data scientifically pure in view of a lot that's been said.

COLLIER   I think it's apparent from many of the things that
go on with respect to environmental issues that people participate in
what's called a risk perception and a risk assessment. What we perceive
to be a threat is often totally disproportionate to the risks involved. There
are numerous examples of this. One is food irradiation. People don't
accept it as a technology because they have a basic fear of radiation. One
recent example was a bill introduced in the Pennsylvania state legislature
to slaughter the horses and sheep in Pennsylvania to reduce methane
output by these animals because of the greenhouse effect. They thought
they'd let the cows off this year, but they weren't sure of next year.

Our perception of risk is directly related to our ability to control
something. When we get into a car, we believe we're in full control of
things. We're willing to accept a relatively huge risk by driving onto a
highway. Yet we're scared to death of things that we can't see or really
don't understand and, therefore, don't have any feeling of control over.
A good part of this debate about the environmental risk is really
environmental perception, and it's based on our inability to understand
and, therefore, to control things.

BENNETT   I agree with Bob. We were discussing that last night. People tend to take a moral perception of environmental protection and say things like, "Look what we're doing to the earth, our mother." While I do believe there's a spiritual tie between heaven, earth, and humans, and that respect for the environment is also respect for God and for ourselves, the fossil history can show us, as Bob noted, populations and environment don't remain the same. A series of volcanic eruptions and/or asteroids may have wiped out the dinosaurs. Within our own lifetime, as human miseries are pointed out, a single volcanic explosion affected world climate. We need to learn from our own past and that of the planet that extinctions tend to come quickly. Things can run away environmentally.

There are various theories about glacial advance or changes in the gases in the environment, in the atmosphere. If not caused by human activities, it can be accelerated or altered in other ways. World population as a whole may be able to get up to a carrying capacity of some billion. But in certain areas -- Amazonia or Africa or the Caribbean -- the population has already had a dramatic environmental effect. People are dying in some areas. They're making a drastic impact on sensitive environmental areas, and none of these things can be entirely divorced from the planet ecosystem. Jean Robert said yesterday that we may be witnessing someone going over a cliff and that's okay so long as it's not us. But, as he said, we're all chained together in complicated ways, and we need to pay attention and make sure that we don't go the way of the dinosaurs.

BRUNGS   Before the session ends I'd like to mention a couple of intellectual changes. One is that we're doing two new kinds of science. When I was trained as a scientist, we, generally speaking, did hands-on experimentation. Now we have computer simulation, the issue that Ernie Spittler raised. We're modeling now instead of experimenting in the laboratory where we at least assumed we had control of the variables. We don't have control over the variables with simulation. We have a poor understanding of this; this type of science is still in its infancy. It will be a significant advance in science, I think, and it's the only way we can scienctifically treat extremely complex systems, but it is not a good science yet. Another kind of science is science by press conference. There's a lot of that lately. It's occurring at the same time we do theology by press conference. In the old days we worked this out with our peers to knock the rough edges off our material. Now the rough edges don't get knocked off anywhere. They get exaggerated. That's another component of what's happening.

Both of these things are happening at a time when the American culture particularly, but I suspect worldwide, is becoming much more numerically illiterate. There is a terrible lack of a sense of quantification. We had a meeting here a few months after Three Mile Island. One of the speakers, Rustum Roy of Penn State, was on Thornburgh's panel of experts immediately after Three Mile Island. He told us that, when he got back to Penn State where he was teaching, the seniors and graduates students he had were terribly worried about this radiation from Three Mile Island.

So he asked them, "How is it going to get here? Is it going to come in on creepy-crawlies or what?" Well, they didn't know how it would get there, but it was a lot of radiation -- 1120. So he asked, "Is that a big number?" "Yeah, 1120 is a big number." He said, "1120 what?" Well, it was 1120 picocuries. So he asked them, "Which is bigger, a megacurie or a picocurie?" He said that he did this, because the only way we get truth in a democracy is by having a vote. Twenty-seven voted for picocuries being the bigger and 22 voted for megacuries being the bigger. Picocuries won. So he spent the rest of the class explaining the 18 orders of magnitude as the difference between the distance to the sun to the diameter of a human hair. He said, "They finally caught it."

This is a general problem. I was reading a headline in the *St. Louis Post Dispatch* the other day that such-and-such a thing (I forget what the topic was) has tripled in the last two years. Wow, doesn't that sound bad? Well, what was the base? Was it from 1 to 3, or was it from 1,000,000 to 3,000,000? Nowhere in the article did it say what the base was. What was I able to draw from that article? Absolutely nothing. I have no idea of the numerical base that they were using.

All these things enter into at least our understanding and approach to some very significant problems. Until we can get a grasp of a notion of quantification, we're lost in this area. We're completely spinning our wheels and not getting anywhere. If we do adopt a specific course, it's just as liable to be in the wrong direction as in the right direction.

GENERAL DISCUSSION

SESSION III

CROSS   In his last remarks Bob Brungs expressed concern about the state of science education. It's a theme we hear a lot about now. We are beset by problems and the citizenry is not equipped to deal with them. Paradoxically, those who try to keep themselves informed are learning that they know more and more about less and less. The more we keep abreast of scientific developments, the more problems we see, the more potential dangers to the planet and to civilization we can identify.

There's an economic problem connected with that. If I'm a scientist acquiring knowledge which opens up vast uncertainties, I'm motivated to let everybody know about it and about my need for more money to investigate this potential danger. That's a situation we face and apparently will continue to face -- the more we know the more we're going to know we don't know. The big challenge to the community at large and the scientific community in particular is to create some mechanism for prioritizing or estimating the relative danger so that we can proceed in some rational manner rather than saying, "Well, we're faced with so many dangers that we can't possibly do anything and let's just go ahead."

BERTRAM   Some years ago in this gathering an essayist advanced a thesis like John's. Perhaps it was a bit more ambitious. It went like this: As we increase our knowledge of the cosmos we find that there is less that we can do with what we know. Moreover, as we increase our control over our environment, we are more aware of how out of control we are. Maybe knowledge is itself a kind of control. Even if we went beyond what we know about the objective situation -- insofar as we can manipulate or control things about us -- the implicit assumption is that, as we come into control as human agents of creation, our responsibility increases. As our responsibility for controlling it increases and we find that we can't live up to the new responsibilities, we are fraught with temptations in several directions. One direction might be that of despair and cynicism, a kind of inevitability. Or we might be tempted to a shrill denial of our lack of control. We might turn toward moralism, a kind of pharisaic pretentiousness that we can't possibly live up to. I thought the drift of some of the statements earlier this morning were heading us in the direction of asking what this implies for our moral and spiritual and theological culture, what we're learning from the sciences.

FORD   I was rereading John Griesbach's impressive treatise that posits a kind of Lockean notion of property rights that had been formed in this country in its first generations and has now been replaced by a new positivist view. He states that that has undermined the Lockean world view. It has resulted in a more complex set up. He then lists some of the pathologies present under the current world view. He cites, for example, legislative stalemate or litigation expenditures on investigation and enforcement. The ultimate shortcoming that he describes is a growing notion that we have a kind of right to use, let's say, common property and common environmental property without a corresponding responsibility for its care or for the consequences thereof. That has somehow come along with the new positivist view because the positivist view, as he describes it, is a strictly utilitarian view based on individual wants that have no more accountability than that. The Lockean notion, despite its own variety of shortcomings, had a long view of the responsibility.

In my own mind I was comparing that with the theology of Martin Luther. If I were to oversimplify and almost caricature some of his views on society, I would say that his view of legal enforcements was a very limited one. He viewed the law as something that could restrain evil in that it could set limits on evil. When asked how one is to promote the good, Luther -- I'm really caricaturing here -- didn't look to the law. He didn't look to the government. He looked to the individual who really understood the message of his own salvation. Luther has a little treatise on what the true Christian prince might be. If people were lucky, they might benefit from the good results coming from such an individual.

At any rate, as Professor Griesbach did at the end of his paper, we can return to the notion of people -- maybe a small number of people -- with a sense of responsibility for the environment making a difference rather than resort to the complex legal system that has no long-term vision.

COLLIER   One of the central issues on the environmental front is the starting point from which we approach the discussion. If we assume that the world is a stage on which the human drama unfolds, we have a different perspective from the one yielded by the idea that humanity is merely one of the species inhabiting the earth with no more rights than any other species. I hold to the former; it represents a very different view of the resources available for the development of that human drama.

BERTRAM   Could I ask you please to repeat the alternatives?

COLLIER   The alternatives are looking at the world as a stage upon which the human drama (human salvation) unfolds and looking on humanity as a species with no more rights than any other creature. Some new "religions" are promoting that today. I think that contemporary nature worship and the New Age movement are classic examples of that latter viewpoint.

FORD   I think that Luther would also see the former as the main purpose of existence, and virtually everything relates to that question.

BYERS   Is there not a third alternative? I use a little phrase, creatures in the midst of creation. Humanity is created in the image and likeness of God and we do in fact play out the drama of our salvation against this background. We're part of the backdrop and so we have this whole set of rights and responsibilities with respect to the furry and non-furry critters out there. Isn't there, somewhere between the drama and the improper view that all species are the same -- that there's no distinction between humanity and the lower species -- room for a third alternative which places us in a context in which we work out the drama of salvation with regard to new relationships with the rest of creation?

FORD   A Franciscan vision. . .

BERTRAM   The great temptation, of which I'm made aware by my more activist students when they speak about the environment, is considering that environment includes everything that is not human.

KARLSTROM   Wendell Berry, one of my favorite authors, writes extensively on these value questions. I would recommend him to everybody here. Some consider him the best essayist in America today. He would make the point, using the Bible as his backdrop, that the Bible views the earth as the Lord's property and it views us as stewards of the earth. What responsibilities does this stewardship imply? From the Christian point of view this is a justifiable viewpoint. The earth is not ours. I find the Lockean concept of ownership to be rather unchristian. Also, from the point of view of an ecologist I find it wrong. I am familiar with the writings of deep ecology and I sympathize with many of them, even some of the biocentric views. I think to some extent we are one of many species and we have to look at ourselves in that regard.

COLLIER   Stewardship is the natural outcome of the position that man is seeking salvation. As a part of that, there's a responsibility, a moral obligation, to take care of that which has been given to us by God. That is an obvious outcome to me.

KARLSTROM   Yes, it's a conditional gift according to Wendell Berry. If we meet those conditions, we can use the fruit. Usufruct is one of the terms used. It's a conditional gift. If we don't use it properly, it's withdrawn. This attitude can fit nicely with the ecological view that if we exceed our carrying capacity, if we foul our nest, we lose the conditional gift. This is something mentioned earlier. Sometimes environmental problems happen rapidly, so we are probably in a state of crisis.

Maybe some of you have seen the <u>Scientific</u> <u>American's</u> issue of September, 1989, "Managing Planet Earth." That title may be a bit arrogant. One of the first authors uses the analogy of bacteria introduced to a nutrient rich Petri dish. The bacteria multiply like crazy, exhaust the resources and then have a massive die-back. The author wonders whether human beings are going to be smart enough to avoid that fate. Are we really smarter than bacteria? It finally comes down to the question of whether we can be good stewards of the gifts that we've been given.

CARTER   I'd like to point out an anomaly that exists when we all are God's creatures together on the same level. We see this vividly illustrated by people who will eat pork chops with pleasure but will object to another person having a fur coat. Why eating a pork chop is acceptable and wearing a fur coat is not is hard for me to figure out. We ought to think clearly about what we mean when we say we're all creatures here together and we all have a kind of equality of Christ. I don't think we really mean it, but we go through the motions of saying that we do.

I would also like to mention this concept of property. We have a concept of property which is different from the concept, let's say, that prevailed in Russia 15 years. I think the Russians have convinced themselves that our concept of property is better than theirs of 15 years ago. Are we about to wiggle-waggle and say, "Well, I know that, but I'm going back to what you guys liked 15 years ago"? I'm simply calling for as much clarity in our thinking as we can possibly attain.

BERTRAM   Quite apart from the options available, are you linking the notion of property, that which is proper to Lee Carter, with the notion that was referred to by earlier speakers, that of responsibility? Are you saying that property that is yours is not just to have and to hold? Are you saying that you can be held accountable.

CARTER   Yes, but there are even problems with that.

Suppose I own a piece of real estate and the city zoning commission says, "That will be great for a shopping center," and I become rich because of that. Suppose also that some other guy has a similar piece of property and the commission says, "That will be great for a park, " and he becomes very poor as a result of that. Many administrative planning decisions are actually allocation of wealth or nonwealth to individuals, and that is a difficult and serious thing. For that reason our guard should be up. I'm not saying we shouldn't ever do it, but our guard should be up against the anomalies of these kinds of situations. They are anomalies and we should carefully try to overcome them. For example, we might say when one guy gets the shopping center and another guy gets the park, the guy that gets the shopping center should make some kind of a payment to the guy that gets the park to equalize it.

WELLMAN   I'd like to reinforce Father Brungs' point about quantification. It seems to me that the unique thing ITEST does is bring together minds willing and able to try to reconcile a field of scientific thought with the range of theological concepts. These are two important spheres. To understand the science and technology we must understand quantification. Unfortunately our understanding of quantification is often terrible. In my own field and in the broader area of, let's say, things nuclear, measuring nuclear radiation and so on, there has been a tendency recently to switch from the curie to the becquerel. As I recall it, there are 37 megabecquerels per millicurie. It's a difficult unit to deal with. When I ask students (having explained what a becquerel is) whether they would prefer to be injected with a hundred curies or a million becquerels dose of a radioactive substance, they will almost always opt for the smaller number, the curies. It's usually the same sort of situation when we have a test. Many of you know about the stress test we do with radioactive thallium. I tell them thallium is a poison and I ask if I can inject these few millicuries of thallium into their arms. They say "No." Then I tell them that they probably eat more stable thallium as a trace element in their diet than I am going to use in this test.

It's difficult for people to understand these concepts. A sense of quantification is basic if one wishes to understand and discuss science and technology in today's world. I worked with the Environmental Protection Agency (EPA) some years ago. I think there is a mortal sin in the field of environmental studies that we ought to be aware of. That mortal sin is that, as soon as somebody is able to create an instrument and detect something and measure it, that detectable amount becomes too much.

This happened in radiation. As soon as you push the forefront ahead in the measuring instrumentation and techniques, that level of presence of a material becomes too much. That, so to speak, is a safe principle in the case of a particular government employee. Then nobody can ever hold them accountable; they set the limits as low as they can. That's a very difficult problem in things environmental. We have to understand things and keep them in perspective, namely, keep the risk assessment a true risk assessment rather than a perception. Most people think a million of anything must be bad, no matter what the units are. I want to suggest a word of caution about all such measurements in the environment. We need to be very careful in enunciating what is a real risk and what is merely a big number that represents something small. We should remain cautious as we discover other environmental problems we don't know about yet. There may be many of them out there now that we don't even know about. They may be a lot worse than anything we suspect.

BLASCHKE   One of the things that prompted me to join this weekend's discussion was the quote in the invitation from Pope John Paul's letter published on January 1, 1990 in which he stated: ". . . education in ecological responsibility is urgent: responsibility for oneself, for others and for the earth. . . . Instead, a true education in responsibility entails a genuine conversion in ways of thought and behavior." Those words puzzled me when I first read them, but I now think he's right. When Dr. Sheahen last night projected his slide showing the consumption of energy in the United States compared with the rest of the world, I had a terrible sense of guilt and said to myself: I must start economizing, turning off the lights and trying to conserve. Yet I find it awful hard to want to do that to help a poor fellow from Nigeria. If Dr. Sheahen's correct in saying that we have to share energy, I don't know exactly how to go about motivating myself, in achieving the genuine conversion in thought and behavior and paying the price that has to be paid. I think we're on the right track with education, with motivation, in getting just ourselves organized to do this in a real sense, to emphasize it.

BERTRAM   This correlates with what you said earlier this morning, John, that, as you hear the reports from knowledgeable people, you're overwhelmed by helplessness even at a micro level. Even at that level you need to look for help.

BLASCHKE   I think I have to change my thinking first be willing to give up some of the luxuries I enjoy.

BERTRAM   I think lots of us are waiting for Thomas Sheahen to discuss the implications of that graph.

SHEAHEN   If you want a comment about that graph, let me remark that that's a report from an oil company. That's one of their charts, and they can show what's happening. In itself it's startling enough.

I wanted to comment on the quote from the Pope saying that education involves a genuine change in behavior. In this country we have over a century or so come to realize that it isn't nice to be cruel to black people, to Jews, Catholics and so on. Yet in all this time the white man's dominance of this continent has continued. We tread into the ground the American Indians who knew how to treat the environment. They took care of it. They were not fanatically against all of the common traces of pollution. The old phrase of the stream that clears itself every ten miles is an Indian notion. A stream needs ten miles for biological purification to take place. The Indians had a balance in which they knew, for example, in contrast to the Lockean concept of property rights, that the earth didn't belong to this or that Indian. It was from God. Neglecting the Indians as a source of knowledge and understanding of the environment is one of our biggest failings. I suggest that, if we are correctly to change our behavior in a way that leads to a better environment, and which from North America we can spread out to all the other countries, we must start by taking decent care of the Indians and making peace with them after what amounts to two or three centuries of oppression. The Pope's words about changing behavior, with particular reference to the environmental issues, demand that we look again at the Indian mentality.

COLLIER   The Indians practiced holistic religion, holistic in its pattern, which they brought over from the East and which is found in Buddhism and other religions forms. It involves a view of man as only a part of nature, not the center of creation. That goes back to what I said previously. It fits into that basic assumption of how we view creation. Christianity has a relationship to the mechanistic views of the world rather than to the holistic. That's one of the tensions that exist between the Eastern and Western religious view of creation.

KARLSTROM   Could you expand on what you mean by the mechanistic view you're talking about?

COLLIER   The mechanistic view is the basis of science, that you can divide creation into understandable subunits -- that the world is a machine composed of working parts. That broad supposition basically drives our thought process about the whole way we approach science.

KARLSTROM   Is that Christian or is it Western culture?

COLLIER   It's Western, but it's bound in many ways to the way the Christian churches have approached the view of creation.

BERTRAM   Would you think it significant, Bob, that that same ancient Eastern heritage to which you trace this holistic creation ethos probably does not have a doctrine of creation, if by that we mean that all that is depends on something other than the creation itself, that it depends on a separate transcendent creator? Would you think that's part of our fault, that we believe that there's a God and then there's what the creator (God) creates?

COLLIER   One very big difference is Christianity's stress on salvation versus the Eastern stress on enlightenment, a process that one goes through in different lives. In some ways the concept of reincarnation is a person's journey through the process of enlightenment. That's totally different. Enlightenment is something that could be self-directed and achieved on one's own. We cannot achieve salvation on our own.

BERTRAM   I had reference to our discussion, starting last evening. We've used Christian or Jewish-Christian or Moslem language like creation, or creator by implication, which is a rhetoric that an Easterner wouldn't use.

BENNETT   I had a question for John Blaschke regarding motivation, although I hate to break up the discussion on perceptions of the world and religion. You brought up a deep issue, John, when you asked how one could find the motivation for a change in behavior. Are you thinking of a theological approach and perceptions of your behavior, or are you thinking more of a social approach? For instance, would you accept the government imposing a gas tax, say, for mass transit? Or would you prefer an individual decision to change your lifestyle and burn less fuel?

BLASCHKE   It's easy to sit around in a room like this and discuss all sorts of wonderful ways to change the world and society. It's difficult in the midst of our plenty to share energy in the sense that Dr. Sheahen's graph implied. I was commenting on the difficulties of figuring out a way to do it and join others in doing it. I think it's almost unattainable in our society. We are selfish and self-centered. In a lot of ways, we're narcissistic and hedonistic. Look at the magazines and stuff we read about luxuries, self-gratification and the joys and pleasures that we're driven to and by. All of these things worry me. That's what's motivating my comments.

CURRIE   Our discussion emphasizes the importance of a two-pronged approach to environmental responsibility or motivation. We talked about both the importance of scientistic understanding and of achieving a new relationship between human beings and creation. In both those approaches we talked about severe limitations -- the difficulty of using scientific models and of achieving some kind of broad-based understanding. Both approaches have their risks. Attempts to achieve some kind of scientific understanding risk a certain amount of trendiness and exaggeration of the crisis. There are various kinds of scare tactics. When people try to achieve a new understanding, they risk the extremes of some of the New Age spirituality.

We seem to be stressing the importance of treating both aspects at once and emphasizing the importance of what we're doing around this table -- trying to bring about some kind of a mutuality between science, theology and the environment. It's difficult because in both cases we're dealing with a very complex set of issues.

LEGUEY-FEILLEUX   I wanted to respond to John Blaschke's comment. I appreciate the difficulty of changing these situations. Often we want to change it all at once. Then we get overwhelmed and fail. We should bear in mind the importance of incremental change. Incremental change is not psychologically rewarding because initially we don't see much of a result. But over the long run incremental change produces results. If we are overwhelmed, especially when it comes to changing things we are doing ourselves, it's probably easier to change a bit at a time. Let's change what we can change right now. Tomorrow we'll find other things we can change. Often enough we try to change drastically, lose our motivation, tire of the effort and revert to our old ways.

MC LEOD   I'd like to follow up on what Bob Collier was presenting. Christians look upon the relationship of individuals to the world in light of the fact that they're made in the image of God. Also there are different traditions in Christianity, both theologians and scriptures scholars interpret the image of God in different ways. Some say that the image, because God is spiritual, must be in the spiritual nature. They put it in the mind and tie it to the notion of creative ordering. They would look upon the earth as desacralized and say that we can use the earth for our own advantage, even to the point of abusing it.

Another Christian tradition came out of Antioch. The Antiochene tradition says that we're the image of God in both body and spirit. There's a profound difference in that kind of attitude. In this tradition the human being is the bond of the universe. We're made up of both matter and of spirit. The angelic world will find its salvation through us as will the material world. In that view Christ, when He became human, took on our salvation. Further, he will restore all things, both the spiritual and the material world. The material world will share in the final glory. Our role is not only merely of being stewards but also to realize the world is part of us, even on an ontological basis.

While we should be sensitive to that bond -- I think the Orientals have the sense that the earth is sacred -- we have the sense that the earth is outside of us, it's different from us. We may occasionally feel one with a beautiful sunset, but we don't share the notion of the American Indians or some of the Oriental religions, Taoism or Shintoism, that there's a spirit world and that we're part of that world. As Christians we may need to rediscover the Antiochene tradition. We need a sense of creation as not merely something to be used. We need a sense that we're one with nature. Then we'd be much more respectful of it.

BERTRAM   Fred, would you see that extending to a salvific affinity with such parts of nature as the cancer cells in my body which are metastasizing and replacing the rest of me, or with the mosquitoes that bring malaria, all God given I presume?

MC LEOD   That gets us into the problem of evil and of good and whether evil is only a lack of the good. I accept the view that God has not created a perfect world, just as he's not made the human being to be perfect. We have a call to create ourselves, to become what we were meant to be. We have also a call to help the world overcome its imperfections. Because the world is not perfect, there will be times when that imperfection will show in disastrous sorts of ways. I think the Lord's desire for us to use our creativity not only for ourselves but also for nature, to help nature become more perfect. It's like a seed which needs to be developed and nurtured. We must direct the creative development or evolution of the world to move forward in a beneficial rather than destructive way. Mosquitoes I think would be a sign of the imperfections within the world, a lack of perfection.

BYERS   I'd like to comment on the relationship between Christianity and science. There's been a lot of work done on that, particularly by a Benedictine, Stanley Jaki. He has spent the better part of a long career demonstrating the relationship between Christianity and the rise of science. Science did not arise in the Eastern world. His argument is that Christianity presents a world that's intelligible. We can't develop a scholastic or philosophical tradition unless we believe that the world is understandable. That leads to a scientific perspective where we take things apart to see how they work.

We are "activists" because we have a creator. That's how we start. We make things. Then there's the traditional and biblical tradition of subduing and dominating creation. No matter how you interpret "dominate," it means to take care of. We have that strong tradition. Christianity is an activist religion that works with things as the drama of salvation unfolds. Eastern religions are essentially passive. They just go with a sense of harmony and Confucian karma: "respect it but don't mess with it." That's an entirely different way of approaching one's place in the world.

The conversion that John Blaschke mentioned interested me. Most of my work for the bishops has to do with evangelization, that is, spreading the Gospel. Our task at this point in the development of that whole move-ment is to try to clarify what the kingdom of God means. Christ came to us and died for us to establish the kingdom of God. But it remains hidden; it is something which is there but which has to emerge through our efforts. What does that mean in practice? Take as an example Tom's graph about R&D consumption. The U.S. consumes a great deal while the Third World countries consume very little. That gross inequality seems rather clearly to be contrary to the kingdom. Why should all these millions of people in the world be living in a state of misery? In that particular case, I think, the growth of the kingdom demands at least doing something about the low end of the scale. So what does conversion mean? Conversion means, in this case at least, bringing people to the point where they can see with the eyes of Christ in this regard and change their behavior accordingly. I certainly agree that motivation changes behavior. Incremental change is needed. The kingdom is not going to come unless we change ourselves.

FORD   There are two contrasting views of nature within Christianity. Representatives of these traditions might be Francis of Assisi and Martin Luther. Bob Bertram once wrote that we can hardly understand the thought of Martin Luther today. When he was traveling through the Alps on his way south he found them terrifying. Bob commented that perhaps we're so imbued with nature that we can't understand Luther's terror of it. Take the notion that contact with nature has deep restorative power because its magnificence inspires peace and serenity. The Pope stated that and it's a notion to which I can respond. But, if I understand correctly, Luther didn't share that at all. I'm not sure what conclusions I'm drawing from that, except perhaps the following.

As I understand it, Luther held that the environment would not benefit by our direct attention to it, but rather attention in other ways. It would perhaps benefit from our indirect attention to it, our attention to what for him was a primary question, namely, that of salvation. Through that the whole of creation perhaps could come more directly into order.

BERTRAM   I should just insert a postscript. The historians from whom I swiped that insight did not locate that perception in Martin Luther exclusively. On the contrary they were contrasting an entire age with our age. It was characteristic of a millennium of Christian thought. By contrast, the sort of <u>geist</u> that developed, particularly with the age of romanticism, sentimentalized nature. We see nature as something quite approachable and unified with us.

SCHNEIDER   My first comment goes back to John Blaschke's noting how difficult motivation is to come by. I've often seen the kind of charts Tom Sheahen used. How often, if ever, do such charts, which point out the inequities of the world, motivate us to action. Are we ever motivated to action simply at the intellectual level? If it frequently happened, the world would be a quite different place. It seems that we are motivated at other levels. One would be a crisis or a perceived crisis. That probably goes back in part to some of Eric's comments this morning, that, even if we're mistaken in some of our concrete judgments about the extent of the environmental crisis, the very attention we give it can produce some very salutary effects.

Another level that I think motivates change or conversion is interpersonal contact. I don't know whether it's true for all people, but it is for me. There's something about knowing real people by real names and faces and coming to understand their situation. That's very different from reading abstract statistics. I don't know how we can make that concept more real on a large scale. Educational institutions can certainly do their part. Religious leaders -- or leaders in any sector -- who understand the problems we face can play a role.

The final thing that might motivate to action is what I call perceived unity with the sectors that are affected. It strikes me that that overlaps into the different ways of perceiving humanity in relationship to the larger natural world. It seems that we have today all the ingredients for a fresh approach in an age that has only within recent decades come to understand that all living forms are unified at the genetic level, have the same common building blocks. As chemists and biologists we can understand the exchange of atomic material. We can take that scientific knowledge as part of a new cosmology, of a new spirituality and of a new basis for understanding our interaction with the rest of creation.

BERTRAM   On that last point, Sister Maxyne, would you go so far as to suggest -- and I certainly am not suggesting this -- that maybe in developing this new Christian cosmology/spirituality Christianity might, if not take a page from some of the new age spiritualities, perhaps steal a march on them by grappling with some of the concerns they're raising?

T. QUINN   Twenty years ago one of our children experienced an emotional problem that, thank God, has been solved. In the course of identification and solution, I became aware of a principle that's supposedly common to the psychological and psychiatric world. It states that the normal human being, whoever that is, goes through the three stages of dependence, independence and interdependence in the development of his or her mentality. I hope that's a valid principle because, if it isn't, I've made many decisions over the last 20 years based on an invalid premise. If this is true for an individual, maybe it's true for the world community, because we're all persons. There is movement through these phases in the development of technology and in the attitudes toward the environment in different segments of the human community. We're not all at the same place, but we are moving through this kind of a situation.

This leads me in one way to respond to John's comments about motivation. I tend to look at an objective as not being change but reaching a balance in whatever the issue is. I have little patience with people who demand change as an objective. I may not be able to quantify balance, but I don't think anyone can quantify change. When I become aware of environmental issues, I respond not so much with a strident attitude demanding change. Rather I ask what can be done to bring things closer to a level of balance. I think change is good. But I am also aware that there are a lot of things that will happen in the natural environment whether I like it or not.

My life is associated with airplanes. As long as I can remember, the one objective of aircraft design was to build a machine that was stable in its airborne environment. In simplest terms this means that, if the airplane is flying along and the pilot releases the controls and just sits there, the airplane will fly along in the last condition in which the pilot let go of the controls. If that was violated, one was in serious trouble. Today we're finding that, if we want an aircraft to go faster and faster, we reach the point where it may be a good idea to make it unstable. We are now building airplanes in which the design objective is a certain degree of instability. We cannot get metals that will hold up at the speeds we want, at the temperatures we encounter, unless we do this. That means that the pilot has to control the airplane at all times. Since he's not capable of doing it, we end up with pilots driving computers, not airplanes. We cannot get to any of these things without experimentation and quantification which I think is part of the premise that Father Brungs raised earlier this morning. I sometimes wonder whether we're beyond being able to reach a balance between computer modelling and hands-on experimentation in many areas of science and technology.

KARLSTROM   Let me go back to that point that Sister Maxyne and Bob Collier made. It could be that the Holy Spirit is speaking to us through the earth. I think Christianity can profit from looking at some of the ideas of Eastern religion, which is that all of matter and all of nature is informed with spirit. The new science, new physics and chemistry shows a deep interdependence in a way that we as a culture and we as Christians had not perceived until the present time. I think that the new scientific vision and a religious vision are quite compatible. There's a book by Thomas Berry called The Dream of the Earth. I would recommend it to everyone. He talks about the new creation story, namely, the scientific view of the earth. The earth should be our main curriculum and our main teacher, since it's God's handiwork. I would like to underscore that point. I think we can and should be open to these ideas and not just consider them to be all pagan and flaky.

WITHERSPOON   One element that I think has not been explicitly treated thus far is that God created the universe and we're gradually discovering many of its mysteries. As was clearly pointed out last night, there are many things in the universe over which we have no control -- hurricanes, volcanos, tornados, lightning and the like, which are extremely destructive. As I understand it, Georgia and the Carolinas are still trying to recover from hurricane Hugo. Therefore, may we not consider that maybe God created the universe and has placed human beings as stewards a universe which in many respects is hostile? Could it not be that people who live in this hostile universe and who grow in the spirit are the persons that God wishes to inhabit the kingdom of heaven? Could it be that those persons who cannot grow in the spirit, who emphasize self and self-protection, are those who are not inhabitants of the kingdom of heaven?

Sister Maxyne raised the question of personal relationship. The personal environment is probably more hostile than the material environment. Think of some of the good times that we have had, but also of some of the very difficult times that have been created by people with whom we are in contact. God created not only the material universe but the people universe, too. Could it be that those people who can grow in the spirit in the midst of a hostile universe of people, even to a greater extent than the hostile universe of the material, could be the inhabitants of the kingdom of heaven?

Therefore, we strive valiantly (or sometimes helplessly) to try to be good stewards of the material universe. But we might at the same time be violently destructive of the personal universe through our attitudes to other people. I suggest that in our discussion of this weekend, we should consider the human environment as well as the material environment as hostile environments whose hostility we might gradually overcome through a growth in the spirit God has given us.

COLLIER   I think that the basic reason that Christians have such a different view of creation and the environment is because of our belief that the soul is separate from the body and does not belong to the world. We don't see ourselves as being part of the world. We're told in

Scripture that we are not of this world. That is a totally different view compared to Eastern thought and seems to me to be the basic reason for this tension between how we view our body and its eventual death and where we will be a thousand years from now.

BRUNGS   I was going to say pretty much the opposite of what you just said, Bob. I think we misunderstand Christianity to a degree at least. We don't realize that it's an urban religion. It's the only major religion I know of that was started in a city. When the Spirit came down upon the apostles after the resurrection, they were sent out into a city. Christianity is a street religion. Even as far back, Bob, as the year 150, St. Irenaeus wrote and preached that the soul is not the person -- the soul <u>and</u> body are. St. Thomas Aquinas says that "even if the soul should attain salvation, I don't." I have to have my body to be me. So there really isn't that dichotomy. This has been sloppy teaching for hundreds of years, part of the legacy, I think, of Rene Descartes. Body and soul are not two things. They're two principles of <u>one</u> thing; they're two aspects of our <u>personal</u> being. There has been a lot of confusion because that teaching has been so sloppily handled. The catechesis has been terrible.

I remember at one of our very early ITEST meetings, Jose Delgado, who made his name and fortune implanting electrodes in people's brains and stimulating certain sensations, swore that, when he did this work, he didn't touch the soul. He touched the person; therefore, he touched the soul as well. For him the soul was one little box and the body was another little box. He put them next to each other. Maybe he could even put one inside the other, but they were still separate. We're not made that way. We're a unity.

We have to use words, both scientific and theological words, carefully. We have to use the concepts even more carefully. This is not done in our society where, to me anyway, the normal course of thought is just simply Jello. It's amorphous and mushy; it has no hard edges anywhere.

I was amused and intrigued by Charlie Ford's mention of Martin Luther. Many of us have been in Europe. If we ride through Europe on the train, especially in mountains, we see villages huddled together with the fields further out. Obviously people came together for protection. There was a need for this kind of a protection, a need that we don't have in this country. We come out of another tradition. In a lot of ways, the American ethos comes out of the Enlightenment. Our extolling of the wilderness as the place of virtue and the city as the cesspool of evil is in a way anti-Christian. I'm painting very broad pictures of reality. But there are places in James Fenimore Cooper's <u>Leatherstocking Saga</u> where this beauty of the wilderness is extolled because the people there were noble and pure whereas back in the city they were all corrupt. That has entered into our psyche and into the American myth.

Europe I believe developed the opposite way. Safety was in the village. Those wolves roaming the mountains were hungry, and a hungry wolf will eat you if he catches you. In those days it probably took a week to cross the Alps on foot and I can see how that might well be terrifying. If one walked across the Rockies on a journey, I think he or she could be frightened. I am. I've been terrified by lightning storms at 13,000 feet where there was nowhere to hide. You hunker down in the rocks and every time you hear thunder, you figured another flash of lightening had missed. And you suffocate with ozone. The mountains have a very terrifying side.

One of the result of our technology is we don't see the fierce side of nature as much. We build a technology to protect ourselves from that fear and danger. Now we're in the course of building a technology to protect ourselves from that technology. I think the idea that nature is totally benign is a less than desirable product of technology. Once we arrive at that idea it's easy to romanticize and sentimentalize nature. That's my quarrel with Walt Disney. <u>Bambi</u> is not my favorite movie.

BERTRAM   I must confess I was tempted to make Robert Brungs' speech. I am glad that it was a Roman Catholic who said what he did about soul and body. The Apostle Paul, for example, when he uses the Greek term <u>soma</u> -- like our word somatic for body -- he uses it to designate the person. It refers to the whole person. The body/soul dichotomy is certainly post-biblical.

GENERAL DISCUSSION

SESSION IV

BERTRAM   I was gratified that the theological questions raised and answered by and large came from the non-clergy. Usually it's the other way around. Having said that, I would like to make a suggestion which you may ignore or honor as you see fit. We will be dealing formally with the theological issues in the spring of 1991. At present we don't want to lose the benefits of the expertise we have available from our scientists and technologists. I would not like to see the contributions that they have already made and will make lost from sight.

BYERS   I realized this past summer, while I was working on a statement on evangelization, that it is the task of the laity to work on these issues. If we leave it to the clergy, they will "ecclesialize" it and make it into a program. We won't convert anybody that way. I originally meant to respond to the dualism implicit in Bob Collier's remarks. I don't want to add to what was said previously except to point out that we are monists, not dualists, in the kingdom of God. Reference to the kingdom of God which is to emerge on earth makes no sense if we take a dualist attitude towards soul and body, if the body is left behind and the soul alone perdures. Then kingdom of God would refer to some pie-in-the-sky-by-and-by. Actually we have both feet firmly planted in this world on a conveyor belt, if you will, that's going to carry us inevitably into the next. We have a job in both spheres so that bringing together of body and spirit is the necessary concomitant of there being such a thing as the kingdom of God.

Sr. Marianne mentioned to me that theology is next March's task. We need to talk about things like waste management and practical stuff. As one who works with the bishops, my notion of dealing with the environment is to convert everyone to Christ. Then everyone is going to be less persnickety and we'll come up with a program. My job is not to work out what's going to happen after people are converted. That's the job of the Holy Spirit. I'll leave the practical suggestions to the scientists.

BLASCHKE   Let me ask David Byers if he could mention this item to the bishops. In the human genome project, plans are afoot to map the DNA sequencing of about 40,000 genes that each of us has and which have to do with what we are, what we will be, what we can be. The costs of trying to map the DNA sequencing of the human genome will be high. Currently the Department of Agriculture, I think, kicks in about $20 million a year, the Department of Energy $50 million, and the Health and Human Services $39 million this year for mapping the human genome. The ostensible purpose of this mapping -- I'm not opposed to it -- is to try to find a better way of controlling genetic diseases.

The problem is what we'll do with this data if we succeed in mapping the human genome. It's entirely possible that within the next 10 to 15 years we'll have routine genome testing of almost every pregnancy. If someone has the genes for a disease, some critical questions will be raised for the parents. It's theoretically possible that if a new-born didn't have the genes for a super brain or a super muscle development he or she might not be totally accepted. What would the Conference of Bishops think about this project and what we ought to do with it?

BYERS   The conference now is not even aware of it, and it wouldn't be easy to make them aware of it. The bishops are not scientifically trained and it's difficult to get them to focus on scientific issues. Any technology can be used for good or for ill. If the genome is mapped -- I think it's projected to take 15 to 20 years -- that technique and data could be used for perfectly good purposes or for bad. It might be used for eugenic purposes. The bishops, I suppose, if they ever address it at all, would say to use it well. I don't know what else they can say.

CRUMP   I'd like to ask the question of anybody here assembled, what is the human genome? Each of us is a unique individual, and whose is the genome? Who decides?

BLASCHKE   It's a collective term for all humanity. Genetic structures and gene arrangements vary. Take, for example, the HLA gene. This is a gene with about 150 different variations on the HLA antigen alone. There are different combinations of arrangements. If we can figure out what the average, common DNA sequencing is in one gene, we can make predictions. Forty thousand is the stated figure of the average number of genes in our 46 chromosomes. It's a little scary to be able to map that and to intervene genetically.

CRUMP   We talked this morning about the uncertainty of climatic information. Even if we had some kind of an overall general picture, we wouldn't know how different variations would work together.

BLASCHKE   I agree. There is a group of diseases associated with HLA. Yet many people with that HLA gene do not have the disease. We don't understand that. Generally genetic diseases are that way. There are a few specific diseases such as cystic fibrosis and Down's syndrome where the gene is absolutely programmed for that disease. There are some diseases which we can predict and into which we might conceivably intervene. That kind of intervention is sort of mind boggling. If we have a child or an embryo, with a gene for cystic fibrosis and we plan to intervene by substituting a healthy gene, we're playing with creation. I can't think beyond that point or even to that point. But it's in the cards.

CRUMP   We already have a hold on many monogenetic diseases without the human genome project. When we get into polygenetic characteristics, only God knows what's there.

BLASCHKE   I introduced the subject mainly to bring in a consideration of the human environment in the next 10, 15 or 20 years. That's going to be a critical question as we go forward.

BRUNGS   I wonder if we'll be able, John, to come up with more than we have now. We're told that our cholesterol ought to be in a certain numerical range. None of us matches that. Or our temperature ought to be 98.6°. Mine is normally 97.5°. We're all individual chemical machines. Each of us is different from everyone else. I accept the notion of general categorizations, so long as we remember that's what they are. Could one of our biologists explain to us what the advantage of this is?

KANE   Sequencing aspects of the disease from the angle of correcting it gives a lot of information. While we may know there's something in a particular chromosome, we have no capability of altering it. What the letters are in a particular piece of DNA dictates the state of that individual with respect to that trait. We then would be able to use the technology in a reset of the letters and insert it specifically into the right place in the chromosome. The technology, in developing those sequences, would provide us with that kind of additional treatment for people who are sick with some genetic disease.

Further down the road there could be some troublesome results. Perhaps, somewhere in the future, instead of passing a medical test to get hired, one might be given a genetic screening test on whether he or she might be genetically predisposed to having a heart attack or some disease or other. The results of these tests might impact on whether an employer would hire him or her or whether insurers would issue a policy or a promotion would be given, and so on. There are other questions about using this sequencing. But the sequencing itself, if you think about it strictly from the medical aspect, would give us a lot more information than just knowing where a particular gene is on a chromosome.

SHERMAN   I'd like to supplement what Jim said. If we find
the sequence, then we can find out what protein product is called for by
that sequence. While in some cases we might not be able to change the
basic hereditary tendencies, we might be able to find out more about the
mechanism of that particular disease and from that point on be able to
figure out a therapy which might alleviate some of the symptoms. I'm
thinking about diseases like cystic fibrosis, Huntington's chorea or one of
the other types of hereditary illness. Sickle cell anemia would be another
good example. These studies open the door to further investigation and
knowledge about treatment and therapy.

I see the danger in insurance companies, employers and so on discrimi-
nating against people whom they have checked for genetic diseases. Also,
I see parents under tremendous pressure to abort the child; it worries me
that parents would be under this pressure. The same thing has happened
with ultrasound. Almost no OB/Gyn doctor today will accept a patient
unless she agrees to have ultrasound examinations throughout the
pregnancy because no doctor wants to be open for a lawsuit for not
having done it. They fear that if the baby is born with a particular
problem, the parents might say, "Well, you didn't do an ultrasound.
Therefore, we're going to sue you." So most of them insist on ultrasound
examinations which apparently are perfectly harmless and give a lot of
information. That's where technology comes in. Then pretty soon, it's
demanded that one use the technology. That's where I see a big danger.

FORD   Since what Marie Sherman mentioned seems to
be related to the law, I would ask John Griesbach to respond. Is there
something about our view of the law that promotes or facilitates the
situation Marie was just describing?

GRIESBACH   I'm actually not aware of, for example, tort
actions against physicians for failure to compel ultrasound. There may be
tort actions for failure to inform parents that ultrasounds are available.

SHERMAN   Perhaps that's what I'm thinking about.

NIEMIRA   There are some cases for failing to advise on the
option of abortion and on ultrasound.

GRIESBACH   I'm not aware of those. Verdicts are awarded for
failure to disclose?

NIEMIRA   For failure to disclose that they had the option to abort this particular child. It's a wrongful birth type thing.

GRIESBACH   I've heard of wrongful birth, but not that kind. It strikes me that that would generally be regarded as understood information. But once we get into the gene study, it's very difficult. It raises many new possibilities, I suppose, for official decisions of the sort that are troubling us. There's likely to be as much confusion and reluctance among the judiciary to look into technical scientific matters as there is with the bishops, and for much the same reason. There's a great deal of uncertainty not only about the mechanism that's at work but its consequences, until there's some experience.

Additionally, there will be the same sort of grappling with what ought to be done. We've already passed the point, if there ever was a realistic shot at it, of foreclosing these kinds of problems by not engaging in the study. That's one way of approaching it, namely, a kind of endorsing and institutionalizing ignorance. Once we gain information, whether in biotechnology or in environmental areas, the next stage is whether or not we're going to push for any kind of official foreclosure of exploiting the technology. If one's going to do that by way of broad prospective rules, we must have some agreement as to what rules maintain and why. We might try to grapple with it on a case-by-case basis, bungling along as we do in most of our common law areas, learning from experience. There's no blueprint in the legal arena that we can apply to this kind of problem.

FORD   Is it conceivable that new information about the genome opens doctors up to new areas of legal vulnerability?

GRIESBACH   Not that I'm aware of, but it depends on what kind and type of treatments are likely to flow from it. From what I can gather from this discussion, that's unknown.

FORD   I mean notions of wrongful birth and other such notions. Do you think the notions exist in the legal realm?

GRIESBACH   Yes. Almost all wrongful birth cases are negligent prenatal treatment cases. And it's called wrongful birth because some of the damages assigned are those connected with the fetus that died. It's a cause of action granted to the fetus, as in wrongful death cases, as against simply malpractice cases brought by the mother or perhaps by the father in a consortium action. It also includes damages that are attributable to injuries to the fetus. The old classic of true wrongful birth cases is a botched up sterilization. There have been botched up abortions, if you want to call it botched up, where damage actions against the physicians have been recognized. But I think there's a nomenclature problem here. The focus is on malpractice within conventional understanding of what appropriate treatments are and a judicial attempt to back away from the question of whether or not this is really a person that is aborted or that is foreclosed with a sterilization.

That's the extent of my understanding of the so-called wrongful birth cases. The damages are almost always nothing because of a long-standing benefits rule. The benefit of having a child outweighs the cost by an uncertain but huge margin. Generally medical expenses are recovered, medical expenses to the mother or the family in having the child after the botched up, let's say, sterilization. Sometimes, if the child is born handicapped, there may be additional future medical costs.

FORD   But you can't envision the kind of situation suggested here, where failure to inform the parents that this child might be born deformed resulted in a lawsuit against that doctor?

GRIESBACH   In failure to inform, yes. Failure to inform of what is discovered on the ultrasound would probably give rise to a cause of action. There may be negligence for failure to inform the parents that ultrasound is available, but I've never seen a case where a physician was held negligent for not personally performing an ultrasound, if he or she had informed the mother that it was available. Maybe Thad Niemira has.

NIEMIRA   I have either read or heard someplace of cases where it was discovered that a child in the womb was defective and claims have been made against the physician for his failure to advise them of the option of an abortion. That's in the back of my mind.

GRIESBACH   That may be right. That may fit within a kind of conventional informed consent cause of action. There is a lot of pressure on physicians to disclose whatever information is available, very little pressure to actually perform activities like abortions or anything else. I presume this would be so in the gene context as well. Who knows what kind of possibilities might come out of this? There's a subtle line between informing and compelling treatment. Down the road we will probably be looking at gene therapy and at gene redesign, what Father Brungs calls genetic enhancement. That raises all kinds of prospects.

R. QUINN   Two years ago I was active in running an organization called COPE in northern Virginia. We worked with women with unwanted pregnancies. I discovered that that's a practice among the obstetricians in our area. I don't know whether it's a law or not. But for a woman over 35 who is pregnant and therefore somewhat more likely than a younger woman to carry a baby with the Down's syndrome, physicians automatically recommend amniocentesis. I don't know whether a sonogram has replaced that or not. They did this to protect themselves from being sued, if she indeed delivered a mentally retarded baby. I don't know how much legality was involved, but it has become a standard practice in the area.

GRIESBACH   Let me get clear on this. This is the physician strongly recommending the amniocentesis?

R. QUINN   Yes. I don't think he could force her to have one.

GRIESBACH   Right. My understanding of the case law in this area is that a physician would be in trouble had he or she not informed the mother that the amniocentesis was available. But the move to the physician strongly encouraging amniocentesis is more likely something going on in the culture rather than coming from the law.

BLASCHKE   In some ways I'm sorry I brought this up. One of the reasons I mentioned it is that I'm addicted to the messages Fr. Brungs has in the ITEST Bulletin. There are some thought-provoking comments in those brief messages. One recent one was that science does not determine its own direction. That had never occurred to me and it alarmed me as I thought about it. This work with the human genome is largely the work of James Watson, the Nobel Prize laureate for the discovery of the structure of DNA. He thinks it's a good idea. Whether or not it comes to fruition ten years from now, it will be a multi-billion dollar project. It may never come to fruition if we have to fund other research for more immediate crises.

Another thing Bob Brungs mentioned in those little notes about three years ago was a quote from an English author, an atheist: "the more we know and the more we learn, the greater the blackness surrounding us becomes." I've thought about a lot since and I've come to believe it's true. No matter what we learn and how far we go, there's a great area of blackness around everything that puzzles and confuses us. The only thing we can do is turn to the Lord in faith to help us around that blackness.

COLLIER   Let's turn the genetic engineering discussion back to the environment from the standpoint of the kinds of things that can be done. One likelihood would be to give endangered species a reprieve because we can store by different methods some ingredient to maintain and amplify stocks of animals and/or plants and prevent their complete disappearance. Instead of narrowing the total gene pool, biotechnology gives promise of expanding it. There's no limit to the numbers of new species that can be created given the possibilities of moving genes from one plant or animal to another. In other words, we won't be destroying biological diversity. We've actually increased its possibility.

There are several ways, as Jim Kane stated, to use genetic engineering technology to attack some of the waste issues. Some of these problems are being approached on a commercial scale with respect to some of the waste coming out of factories. If we can engineer microbes to convert a toxic waste into something useful, it would help. This is being attempted on a commercial scale in some instances. So, there are several ways to use genetic engineering technology to try to improve the environment we live in, and there's also the opportunity to save some endangered species.

KARLSTROM   When you say save endangered species, do you mean resurrect extinct species? Can we get the California condor back?

COLLIER   Some people would like to try that if they could find some intact nuclei. Some frozen mastodons have been found. In my opinion that's nothing more than a game. There's no obvious value to society with something like that.

KARLSTROM   Saving species whose habitat has been destroyed by human activity, such as cutting down the rain forest, would have no purpose either.

COLLIER   We all have to have a habitat to live in. But there are cases of animals coming back to the habitats where they had become extinct. Missouri has been successful in restoring several species. The wild turkey is one and there are several others.

LEGUEY-FEILLEUX   On the issue of the drive for new knowledge, we have to combat the fear of new knowledge. Ultimately, when we move towards the unknown, some of the applications may be perverse. That doesn't make the drive for new knowledge evil. Knowing is not the doing of evil. If knowledge becomes available and it turns out that good things can be done with it, so much the better. If it turns out it's absolutely impossible to do anything good, can we say that that kind of knowledge was not evil in itself? Well, at the time we discovered it we didn't know that it couldn't be applied in a good way. Take nuclear fission or fusion! That nuclear weapons are now possible doesn't invalidate the worth of ongoing nuclear research even if it turned out we could find nothing constructive out of that technology.

There has always been a tendency for people to condemn drives towards new knowledge. "We don't know where it's leading us, so let's not touch it." That has happened at almost every stage of advanced research. We should refrain from that innate fear. We should endeavor to control it, if it turns out we are putting it to an evil use. But to say that we must not push a particular research area any further shortchanges the future.

BENNETT   The pursuit of knowledge is always a noble thing, but there are many avenues to go. If a particular project costs $100 million, may not there be a lot more down to earth things that we could do for health care or energy problems throughout the world?

LEGUEY-FEILLEUX   That's definitely a fair question to ask.

BERTRAM   Jean Robert, let me try to recapitulate what I thought some of the other people were saying earlier. They weren't discussing a kind of unreasoned, know-nothingness for its own sake. I understood people to be saying there's a genuine dilemma here, at least a moral dilemma. While there may be an irrepressible impulse to know what is knowable, there are the realities about human fallenness which raise the question whether the new knowledge can be managed responsibly. That's one side of the moral dilemma. The other is that, if knowledge is available to us, are we not under some kind of mandate to take the risks. Maybe I misrepresent the earlier speakers. I thought the dilemma was more real than your response seemed to indicate. Speaking for myself, I would think that trying to suppress what is knowable --

prescinding from all questions like cost -- ultimately is doomed to fail. Sooner or later that knowledge will win out. By itself, that does not alleviate the danger that people who come by that new knowledge will not be responsible stewards of it. I think that presents a dilemma.

LEGUEY-FEILLEUX   I agree, but I would go one step beyond it. The more advanced we are in our controlling our environment the greater is our responsibility. Therefore, the greater is the need to realize our need consciously to train ourselves in value judgments to a much greater extent than those who came before us. We'd better face that.

BERTRAM   And the greater our guilt if we can't live up to what we now know. Is that true?

LEGUEY-FEILLEUX   Yes, yes.

GRIESBACH   To connect some of what Dan Bennett mentioned with what Jean Robert and Bob Bertram said, we're inveterately curious. That's not going to be purged from us. But we make some sorts of satisfying our curiosity easier than others, primarily by funding them. We fund some areas rather than others. Generally we can distinguish between basic research and applied research. But there's also a distinction, for example, between defense related basic research and biomedical research. Do we need more basic research in environmental control or in mitigation of pollutants? I'm thinking, for example, of a potential for research in such things as ground water control or that sort of thing.

KARLSTROM   There is an English chemist by the name of James Lovelock with whom some of you may be familiar. He came up with the Gaia hypothesis, that the earth itself, the biosphere, acts like a single-celled organism to regulate the conditions for the perpetuation of life. It's a theory that many scientists are taking seriously. In fact, Stephen Schneider at NCAR set up a conference on that hypothesis. Lovelock states that we're in critical need of a new profession of "geophysicians." It's time, he says, to establish a new discipline called "planetary medicine." What are the vital signs of the earth and where are those vital signs giving warning? I think he's right. In what departments that discipline might bloom, I don't know. Perhaps in biology, geology, geography.

LEGUEY-FEILLEUX   There is no doubt that we have a role as scientists to try to curb that tendency to fund the spectacular. Look at the way we have failed to help the Third World because we couldn't bring our technology down to a level they could use. They can't use our fancy technology. They'd be better off if we devised appropriate technology. The United Nations devised the U.N. university without a campus to be able to provide services that are not normally provided. We need more of that. Because it's unspectacular, it's several steps back from what we can provide for people at a different level of advancement. It's time we fund needed projects which don't have the pizazz to generate grants. We can push for funding for such research by the learned societies to which we belong. Dan (Bennett), you were the first one to raise the issue. Is it worth the money we're going to spend, considered all the other needs to be met? Can't we serve this type of need in our research?

FORD   I'm still wondering about Jean Robert's vision of us as scientists proceeding with new knowledge. It's never been completely clear to me what the rationale for that is. From the Christian point of view I don't see its rationale, particularly since more often than not things don't turn out the way we anticipate they will. Richard Rubinstein called this non-achievement of our expectations the "cunning of history."

LEGUEY-FEILLEUX Father Walter Ong addressed that question in different terms about 10 or 15 years ago when he defended the ivory tower. The ivory tower was looked down on. Remember, we ran after the practical, the relevant. He suggested that we fight for the irrelevant and knowledge for its own sake, not knowing where it's going to lead us. It's a great adventure to find out what's there. Most of our results come from research done without knowing where it was leading. That knowledge became available for whatever good purposes could be derived from it.

FORD   Or evil purposes.

LEGUEY-FEILLEUX Yes, but perhaps that's where the much maligned principle of double effect enters the picture. That principle hasn't been made irrelevant because some people misused it. Every human action has a multiplicity of consequences, and we haven't made the good uses evil because simultaneously there is a bad consequence. Take driving a car. That there are 55,000 deaths a year on the highways does not make a car evil or driving it sinful. That's where we enter the picture. We have to be able to sort it out. It gets trickier as our society becomes more complex, especially as it escapes the controls of the scientists. We do our work and then somebody takes it from our hands. That's probably what people are afraid of with Monsanto and genetic research. Their next step would be to ban the research. We have to fight that temptation.

FORD   I'm questioning our ability to sort it out.

LEGUEY-FEILLEUX   In that case, we're no longer morally responsible.

FORD   I wouldn't draw that conclusion. I'm only questioning our ability to sort it out.

LEGUEY-FEILLEUX   If you say we are unable to sort it out, then we can no longer morally engage in the pursuit.

FORD   I didn't say that. I'm talking about problems which seem to be without answers.

LEGUEY-FEILLEUX   Well, as a problematic consideration, yes. So long as you don't take the next step, I have no problem. Perhaps with more exertion, work or deeper insights, we can prevail. It is our duty to do it.

BERTRAM   I thought that what was being implied, if not explicitly stated, is that the dilemma is more insuperable than we at first said. I thought the original implication was that the dilemma is the following: on the one hand, the lust to know -- libido cognoscendi -- is irrepressible in any case. But once the knowledge is out, we face the problem of good stewardship of what we know. Given the unreliability of the human, it could go wrong. What was left unsaid was another dimension to the dilemma which makes it sort of infinite in its regress, namely, what would be the practical alternative. Would we suppress research? In that case, whom would we choose to do that? Would those who suppressed it or controlled it be any more reliable than those humans to whom we would entrust the new knowledge if we had it? It's the sort of dilemma that Plato, for one, raised. I thought that was being implied. There's no practical way out of this dilemma once we introduce practical considerations to it.

LEGUEY-FEILLEUX   After we, say, achieve a new level of knowledge and foresee that some groups of people, if in possession of the knowledge, are likely to use it in a way we'll condemn, then it's our responsibility not to channel that knowledge to their hands. That becomes a practical judgment. If we know that some government agencies are desperately looking for a new way to create a weapon, and if we have moral qualms about that weapon, it becomes our responsibility, to see to it that this knowledge will not reach them. I'm not questioning that.

BERTRAM   Robert Oppenheimer's story is a tragic epic of that very dilemma.

LEGUEY-FEILLEUX   I'm concerned that many of those scientists started worrying about it after the fact. They should have been concerned while they were evolving that new knowledge. Many of them used faulty moral reasoning to assess what they were doing. I can sympathize with that. When we judge ourselves we don't always use the best reasoning.

FORD   When you say "best reasoning" are you implying that they didn't study logic or something like that? Reason wasn't the primary issue. For them it was a much deeper issue than just reasoning.

LEGUEY-FEILLEUX   I didn't mean it in that sense. The nature of conscience is involved. Counsellors often must reassure people, who come with very hurt consciences to be reassured, that they did nothing wrong. But they were so involved in the action, they didn't have a clear picture of what they had done. I think we have to be aware of that. The old axiom is don't judge yourself. It's better to be judged by others who have some distance from the action. It permits a better perspective.

BERTRAM   About 45 minutes ago, with John Blaschke's provocative comment about the human genome project, we switched from the specific theme of this conference (the external environment) to what in our last March meeting might have been called the internal environment. It proved to be, in my judgment, a very fruitful discussion. Somewhere in the course of that discussion, we shifted gears to another theme. That was the question of the stewardship of knowing, i.e., as a way of knowing and the responsibilities and hazards, thereunto appertaining. If I'm correct, we're talking about the responsibility and stewardship of knowing and we're near our original subject of the external environment. Those scientists who explore the internal and external environments have in common the stakes that are at issue in scientific discovery and research. I would encourage a return to the scientific findings of the external environment theme, keeping in mind the moral issues involved, not to mention the survival issues of the planet and its inhabitants.

COLLIER   I want to make a quick comment on the pursuit of knowledge. In agricultural science it's clear what the challenge is. The rate of increase in the population growth absolutely dictates that we must increase our food supply. To keep up with the rate of population change, we have to increase our food supply about 1 percent faster than the rate at which the population grows. For the last 50 years we've conducted what's called agricultural intensification, increasing inputs of fertilizer and energy. We've been successful in this. But that process has now reached its limits, and it's not possible, at least in this country, to increase agricultural input and get a meaningful change in agricultural output. So, the arrival of genetic engineering is absolutely essential to prevent widespread famine in the next 20 years.

There is one way to stop this, of course, and that's to stop population growth. Then we won't need this knowledge to keep up with people's need to eat three times a day. I'll give you one example. Two years ago we had a drought. We slipped below the magic number of 60 day supply of corn in our surplus stores. We've not yet made up that surplus. The impact on the Third World is not that the cost of corn goes up; the availability of corn drops. Our surplus is what's available to give to other countries. If we are below a 60 day supply, we can't be as free as we could be to give to those countries which may be having a famine. We won't make up that critical gap until 1993 at the present rate at which we're producing grain. So it's absolutely essential that we find ways other than increased intensification to increase our agricultural productivity. We can't change the metabolic efficiency of the human, but we can attempt to change the metabolic efficiency of the plant or animal to increase its output. That is essentially what genetic engineering is all about.

CROSS   I'd like to consider science and philosophy. The more radically new the technology is, the more uncertainty connected with it. The less history we have with it, the more caution we should exercise in pursuing it. I could name a number of projects that fit that category -- the linear accelerator, genetic engineering in general, planetary colonization, the development of climate models. The one that impresses me the most, maybe because I'm a psychologist, is the idea of helping the Third World and learning what would really help them. If it's giving them more food which results simply in a population explosion, or if it's selling them our consumer culture, we may not be helping them at all. I think there are very large questions in that area.

The other comment I have is philosophical. I was taught that theology is the queen of the sciences, next philosophy and then the empirical sciences. That hierarchy makes sense to me. We can argue that theology is the queen because she has the revealed truth to add to all the philosophical and scientific information. Philosophy is more speculative, has broader vision and is bolder in what it will speculate about than the empirical sciences and has a claim of superiority over them. I detect in our discussion so far evidence of holism and reductionism, of dualism and idealism, of positivism and Kantian ethics, of pragmatism and instrumentalism. I would like to know more about that.

SCHNEIDER   When Charles (Ford) asked the question that provoked the last round of discussion, I thought he was trying to get at a Christian imperative for knowledge. I don't know that this is an imperative or strictly Christian, but it's probably common to those who believe in a personal god. It seems to me that, if there's anything that comes close to fitting your description, it would be "what you know and see to be beautiful in its greater detail is to be used for praise." Not for a moment do I suggest that the bulk of the scientific world operates out of that motivation, at least consciously. I don't think it frees us from making the decisions that Dan Bennett raised -- what can we afford to do in the light of many competing needs.

FORD   Would you repeat that?

SCHNEIDER   Let me be concrete. In physical chemistry, when I teach about the nature of the atom, I'm always aware of protein synthesis. I teach the smattering of biochemistry I know from two or three years as a student. We are so utterly fortunate to live in an age when we can know and see this beauty. We understand what generations before us -- holy people, poets and psalmists -- didn't know. The psalms and the poetry of the earlier eras praise the beauty of the heavens and of the earth. I can praise God for the structure of the atom or the absolute elegance of protein synthesis or other such things. That for me would be the godly imperative to knowledge.

KARLSTROM   In response to Charles Ford's comment let me remind you of the biblical statement that God confounds the clever. It does seem that often the more we know the more confusing things become. In that sense his question is an excellent one. What we do with the knowledge is a matter of motivation and morality which seems notably lacking in the leaders of the world in this century particularly. Is having greater and more powerful tools a good for humanity? That was Oppenheimer's great dilemma. I don't have an answer for that but let me make an observation. In a Christian journal called <u>Firmament: Christian Ecology</u> an article tells of some native African porters travelling along with some sahib. He was in a hurry and urged greater speed. But the chief porter said, "We have to stop and wait for our souls to catch up to us." We have not stopped this century for our souls catch up to us.

KANE   I want to comment on some these ideas. If we start out knowing what we are after in research, why do the research? Let me give a bit of a history of genetic engineering. Do we really want to fund, in this particular case, the human genome project? In the 1950s some scientists were studying viruses that affect not humans but bacteria. During that particular study, they noted that, when one virus would happen to affect one bacteria, it could not grow. Yet if it somehow managed to finagle through the organism's system and then was taken out and reinfected, it did in fact grow. From that came a very early understanding of what scientists call restriction enzymes. These are basically scissors capable of cutting DNA. The specificity of that particular action was not appreciated until the '70s. That observation in the '50s, in a seemingly irrelevant area of knowledge -- bacterial virology, if you will -- is in effect the cornerstone of the genetic engineering that developed in all the biotechnologies since then. To try to prioritize research is tricky at best and is not very predictive. Some of the other ideas that have come up about ground water regulation and new sciences have in fact come out of this research as well.

Scientists themselves called a halt to genetic engineering in the late '70s and early '80s. There were scientific concerns about what this technology might do. We felt the need for a moratorium in effect to let our souls catch up. For a period of time there was a good deal of regulation on this work. Science was proceeding at a very slow pace until a data base was generated whereby we could begin to make some projections or at least come to some understanding of what may or may not happen. Some things were completely outlawed. We could not, for instance, genetically engineer a known toxin -- something which could be pathogenic to man, plant, or animal -- into an organism for fear it could get out. As the data base was built and regulations began to be relaxed, we knew what we could and couldn't do. There are still rules against trying to work with toxins, trying to work with chemical and biological warfare kinds of protein products. If we're going to do genetic engineering on some particular organism and if it's not on this specific list, it's not a problem.

We've gone through the scientific process, at least to a degree. Some say that genetic engineering is now ready to do some things for the benefit of mankind. How this particular science will impact the environment has generated a number of "geophysicians," if you will, because people are now asking questions about the effect on the environment of a release of a genetically engineered microorganism into it. Say we would like to use a genetically engineered organism to put in the field either to kill pests or reduce the need for fertilizer or whatever the chemical treatment would be! There's always a concern about what that organism will do in the soil. Is it going to get away? Will it create more problems than it solves? Microbial ecology is trying to understand ecosystems and the implications of adding something to them. We're trying to learn what the disruption may or may not be. We've brought up many things here that genetic engineering has followed quite well in its efforts to control itself and understand what we really need to worry about. The problem is trying to translate that to a public. With any radical technology we have a difficult task convincing the public that, so far as we can tell, it is not a problem. We can't be 100 percent sure. We must do what we can and then use our best judgment. Genetic engineering and its potential for things that benefit the environment needs support.

BRUNGS   In the early days of ITEST I would have hesitated to do what I'm going to do now because I didn't want to prejudice the direction of the meeting. After 45 meetings I know I never could have set the direction anyway so I'm no longer worried about that. I'd like to direct three questions to specific essayists for response sometime during the remainder of the meeting. Before I do that, I want to thank you because so far in this meeting nobody has talked about "nucular" anything. I'm sick of hearing the word "nuclear" so badly mis-pronounced.

I talked to Tom Sheahen last night about radioactive waste disposal. I would like to include a discussion about that at some point. Another topic relates to John Griesbach's paper, namely, the rise of omnipresent and almost omnipotent regulatory agencies. What is the future direction of regulation, of the creation of bureaucracies which are radically unanswerable to electorates, making laws to which all the rest of us must conform? That's a major issue in environmental considerations and others. Thirdly, I would be sorry to see Nancy Kete's paper on acid rain entirely lost from the meeting. Ben, could you say something about that issue?

ABELL   I will tomorrow.

BRUNGS   Fine. That's all I had to say and I have no fear
that I have just determined our direction for the rest of the meeting.

GENERAL DISCUSSION

SESSION V

SHEAHEN   Energy sources are one of our greatest environmental concerns. Remember, the whole society runs on electricity. We
have basically two choices for the long haul: either we burn coal or use
nuclear power. If we burn coal, we release CO_2 which may bring on a
greenhouse effect. That's the disadvantage of coal. On the other hand,
nuclear power has some disadvantages. The most worrisome of these is
proliferation, namely, that terrorists will steal radioactive material and
make bombs. The second problem with it is handling the waste. There is
not a widely recognized effective long-term solution to that problem.

We must distinguish between fusion, natural radioactivity, fission
occurring in a standard reactor and fission occurring in a breeder reactor.
In a fusion process, hydrogen goes to helium because of the interchange
of neutrons, protons, and so on. That process occurs in the core of the
sun. We'd like to be able to control that process on earth. The best we've
done is the hydrogen bomb which, of course, is not controlled. The effort
to control fusion has been going on for a long time without much success.

In view of the afternoon's discussion on the moral dilemmas in the
acquisition of certain kinds of information, it's interesting to note that the
problem of building a controlled fusion reactor is the same problem as
developing a better hydrogen bomb. It is an interesting moral statement
for someone to say, "I'm not going to work in that field because it deals
with bombs." Yet that is also the field of controlled fusion which nearly
everybody wants. I've seen this dilemma as a physical scientist. Biologists
are now staring at those questions. In genetic engineering and its
opportunities, we see these swords have two edges. What we've seen in
the physical sciences with bombs and fusion we'll probably see occurring
along the same lines in the biological area.

Anyway, in fusion, hydrogen goes to helium. In a star, the process
continues in the central core by making all the usual stuff like carbon,
nitrogen, oxygen, sulfur, most of the periodic table, up to iron (Fe). It
does not produce any elements heavier than Fe^{56}. Where do the heavier
elements come from? When a star blows up at the end of its life, the
supernova produces all the other stuff which includes things like gold,
lead, the actinides, thorium, uranium. It also produces plutonium,
neptunium, californium and nobelium. All the elements heavier than
uranium decay rapidly. Some of it goes away in seconds, some in weeks
and some in years. At any rate, there hasn't been a supernova in our
immediate area in some billions of years. As a result, the heaviest thing
left is uranium which has a half life of billions of years.

In natural radioactivity, one of the elements in the earth's crust, uranium (U^{235}), decays through the steps of radium, radon gas, and finally stops when it becomes lead (Pb^{208}), which is very stable. In the soil next to your basement there's a little bit of uranium which decays into radon. The radon can get into your basement. After some number of days, the radon breaks down into polonium or something like it and then on down the chain to lead. That's natural radioactivity.

U^{238}, another isotope of uranium is essentially stable. Its half life is so many billions of years that we can more or less count on it being around. The present reactors start with U^{235}. This is a relatively scarce isotope because it has a much shorter half life than U^{238}. It's been decaying since the formation of the planet, so there isn't much U^{235} left. We enrich this, put it in the fuel rod, and hit it with a neutron. In fissioning it breaks almost in half, into fragments, like iodine, xenon, cesium and so on. It also produces more neutrons which then break up other uranium atoms.

If we bring a lot of these atoms together in the same place so that each one breaks apart and sends off two extra neutrons which hit more, we get what's called a critical mass. That produces a bomb. In trying to build a reactor, the number one thing on our list of safety features is never to put enough uranium in the same place to allow the creation of a critical mass. In every step of every process dealing with any kind of reactor, the first question is whether there's any chance of going critical. The answer had better be no. These are the reactors we have. Along the line, they produce waste like americium (Am^{241}), neptunium and other things. Some of this waste has half lives of hundreds of thousands of years.

In a breeder reactor, we work with U^{238} because it is much more plentiful than U^{235}. If we hit U^{238} with a neutron, it goes into plutonium (Pu^{239}) and releases an electron. When Pu^{239} decays, it produces approximately the same waste as U^{235} -- cesium, iodine and other radioactive materials. These too produce americium and neptunium.

This is the storage problem, what's called high level nuclear waste. A spent fuel rod contains other junk like cesium, iodine, xenon, americium, neptunium, and probably 50 or more elements of the periodic table. What are we going to do with this waste? We now keep them at the reactor site in what are called "swimming pools." No one regards that as a long-term solution. The spent rods don't take up much space. A typical swimming pool at a nuclear reactor site is about the size of this room (about 60 x 20 x 8 ft), maybe a little deeper. We put spent fuel rods in the water and every time we change the fuel rods, we put a few more in. In about 80 years we'll fill the swimming pool. This is a short-term solution to the waste problem. We need a permanent storage site for radioactive waste. "Permanent," when you're talking about 10,000 years or so, is a very long time. Nobody knows how to do <u>anything</u> with a 10,000 year horizon. The waste people have a place called Yucca Mountain in Nevada where they would like to bury it. That's being debated in Congress. People in 49 states love that solution; those in Nevada hate it. At any rate, that's the so-called permanent disposal site.

A third possibility is reprocessing the spent fuel rods by putting the radioactive waste back into the reactor. If we put the waste products someplace where there are lots and lots of neutrons, that is, back into the core of the reactor, they will decay very quickly. They won't last 10,000 years. They'll be gone in hours or days. Essentially the word "half life" refers to the rate of radioactive decay of a material "out in the open." If we put something in a reactor, it won't last very long if a lot of neutrons are coming by. One way to get rid of the waste is to keep it in the fuel rod, reprocess it, and put it back into the reactor core. That's the direction in which modern nuclear technology is headed. We have hopes that that process will solve the difficult waste problem. We think that, once that's solved, we've overcome the problems of nuclear technology.

Incidentally, the reprocessed fuel rods are so incredibly radioactive that we can handle them only behind extremely thick glass plates. Nobody in his or her right mind, not even people in their wrong mind, would steal this. No terrorist is so suicidal that he or she would steal this stuff. So we knock off some of the diversion problems (the terrorist problem) and the storage problem at once. That is why in my five "straw-man statements" of last night, I said that nuclear power is the best solution. The environmental problems are being overcome at the present time. As a result, the next generation of nuclear power facilities, I argue, is considerably better than coal plants. For that reason, I think nuclear power has a very bright future and will help solve the greenhouse problem by getting us away from burning coal.

GRIESBACH   How nasty is it stuff once it's gone through a couple of recycling processes? I'm not thinking of terrorists attempting to steal it, but rather attempting to blow up the facility or something like that.

SHEAHEN   Here's the process. If we take something like Am^{241} and hit it hard enough with a few neutrons, it will fission. It's practically the same weight as uranium, so when it splits, the fragments of a nucleus produce a lot of intermediate elements. They become the usual debris that we know how to handle and with a half life of days to a few years or so. Once they are reprocessed and become part of the fuel rod, they are just as eligible as uranium for fissioning.

COLLIER   France is almost totally nuclear. How has the French nuclear industry handled the public safety issue?

SHEAHEN   I don't know for sure. I think that the leading factor in France has been successful education of the public. The French are not as fanatically anti-technology as many Americans.

KARLSTROM   I saw a TV special on this and I believe that it's because the government has subsidized it 1,000 percent. All the nuclear plants they build are of the same model, of the same construction. The nation, the government, has invested heavily in this. As you say, they've educated the people. Some would say they've brainwashed the people. They make the case for nuclear power very convincingly. Not only that, they don't have any other recourse if they want this kind of energy. They don't have fossil fuels.

COLLIER   I'm not aware of their having an accident. Their safety record is good. Do they have safety procedures that others don't?

SHEAHEN   The only bad event in this country was Three Mile Island. It turned out so badly because somebody intervened when he or she should have let it alone. Perhaps the training is better in France. If a control system is any good, it will bring a reactor excursion back under control and shut it off. I presume that the French, when they have such excursions, which do not occur all the time, have not had any mishaps of the scale of Three Mile Island.

BENNETT   What about the Canadian heavy water reactors?

SHEAHEN   That is a process known as the CANDU cycle. It's a good one and it's inherently safe. It shuts off if anything goes wrong. It uses uranium at a pretty fast rate. It doesn't change the reprocessing or storage choice. When they finish with a fuel rod, they must either recycle it or store it.

BENNETT   Which one of the uraniums does it use?

SHEAHEN   I forget. I think it's U^{238} but I'm not sure.

KARLSTROM   Are you familiar with the State of the World books which have come out since '84 from the World Watch Society. Each year there are chapters on energy: creating a sustainable energy future, raising energy efficiency, shifting to renewable energy. Decommissioning nuclear power plants is one. They also mention radioactive waste. We produce about 1400 tons of high level radioactive waste which, as I understand it, we are not disposing of permanently. This waste has a half life of 26,000 years. No civilization that we know of has lasted over 5,000. This puts a burden on many generations to come up with a technological solution that works for that length of time.

I understand from people who've been working on the various test disposal sites, that it's difficult, nay impossible, to guarantee that sort of geologic stability because of faulting, ground water movement, geologic processes that operate at a much faster rate than that. We're talking about high level radioactive materials being produced at the rate of 1400 tons per year. We have no technological solution for that waste disposal at present. You mentioned that they're working on reprocessing it into fuels, but that's not been done, has it?

SHEAHEN   Let me give you a little background on that. In an article I wrote for ITEST last June, I told the story of the integral fast reactor. This is an invention of Argonne National Lab. They've run it for 25 years or so. In the late '80s, the budget was finally approved to build a true full-scale reprocessing plant. That construction is almost complete. This plant will do on a full-scale basis the same things that were done throughout the '70s and early '80s with respect to reprocessing. Fuel rods containing the 26,000 year stuff have already gone back into the reactor and all that waste was burned up. They've reached the point where the scientific and engineering demonstration is complete. Now it's a matter of putting a pilot plan out to show all the world that this will work. Plant construction takes several years, and the reactor probably won't turn on until about 1996. When it does, all the technology we gained during the early '80s will be demonstrated. We're happy with the way this project is going. It solves the real problem you brought up. Since geology changes faster than 26,000 year periods, that kind of permanent storage isn't a good solution. People have been aware of those objections for 20 years and they're well taken. Research has been done toward solving them.

KARLSTROM   In its chapter on decommissioning nuclear power plants World Watch points out these plants have a life expectancy of about 30 years. Although there are about 500 plants now in operation worldwide, none have been decommissioned. This is going to pose challenging engineering problems and may cost an average of a billion dollars a plant to decommission. If we add to the 26,000 years expectancy of the wastes the enormous expense of constructing such plants and then decommissioning them, nuclear power seems to be an impractical solution. The World Watch people in fact say that our best bets for energy are conservation, efficiency in the short run and renewable energy sources in the long run. They make their case, and I think it's worth considering. Energy for most of the earth's biosphere is the sun. That's the allowance of energy. That's our renewable resource. In order to develop a sustainable society we need to live within the renewable realm.

If we continue to burn in the nonrenewable realm, we expend this unrepeatable gift of fossil fuels and uranium which took millions and billions of years to generate. I've heard that in one year humankind consumes the amount of energy that nature took a million years to produce. That's not sustainable. Maybe that's behind the J curve of population growth, another one of our big problems. This rate of energy consumption is a big part of the problem, not the solution. I challenge the assumptions in your paper that high quality of life correlates with high energy use. Jesus, Mozart, Beethoven, Newton and Shakespeare lived perfectly well without electricity, automobiles, word processors or computers. Maybe the quality of their life was no worse than ours. It might have been better, since they had a consumption pattern more appropriate for our species. Buddhist economics says that we should be looking for maximum well being with minimum energy consumption. Maybe minimum consumption should be a primary goal for Christians, because we know that if we consume something, someone else can't.

SHEAHEN   May I break in before I forget some of the points you've made. The first generation of nuclear power plants disappointed everybody because of the problems that have come along. Let me say of the second generation -- the Canadian reactor, the high temperature gas reactor, the integral fast reactor -- are built to overcome obstacles which arose in the first generation reactors. The unhappy legacy of the first round of nuclear power plants is something that the second round deals with. No one is building more of the old types.

Certainly conservation is the best thing we can do as far as energy supply goes. In running a furnace to melt steel, it's stupid to send hot nitrogen up the stack. Yet hot nitrogen is the leading product in tonnage of most steel mills. Anything we can do in the direction of conservation clearly helps. The renewables have been small and it can be calculated that they will continue to be small. I see an important point in some of the things being achieved in genetic engineering and agriculture and places like that. It may well be that we can some day build a bacteria that will do exactly what we want to do in the way of energy conversion.

Converting some kind of agricultural waste to ethanol, to help reduce the use of oil and gasoline in cars may be ahead of us. But the last time anybody did any serious engineering calculations for the renewables, it was only a very small contribution to take to the bank. So for the renewables to be an important contributor will take some advances, I would think, in our understanding of biotechnology.

KARLSTROM   I quote from Wendell Berry's "The Use of Energy": "One possibility is just to tag along with the fantasists in government and industry who would have us believe that we can pursue our ideals of affluence, comfort, mobility, and leisure indefinitely. This curious fate is predicated on the notion that we will soon develop unlimited new sources of energy -- domestic oil fields, shale oil, gasified coal, nuclear power, solar energy and so on. This is fantastical because the basic cause of the energy crisis is not scarcity, it is moral ignorance and weakness of character. We don't know how to use energy or what to use it for, and we cannot restrain ourselves. Our time is characterized as much by the abuse and waste of human energy as it is by the abuse and waste of fossil fuel energy. Nuclear power, if we are to believe its advocates, is presumably going to be well used by the same mentality that has egregiously devalued and misapplied man and woman power. If we had a limited supply of solar or wind power, we would use that destructively, too, for the same reasons. So the issue is not of supply but of use. The energy crisis is not a crisis of technology but of morality. We already have available more power than we so far dared to use. . . . The issue is restraint. The energy crisis reduces to a single question: can we forebear to do anything that we are able to do. Or to put the question in the words of Ivan Illich, can we, believing in the effectiveness of power, see the disproportionately greater effectiveness of abstaining from its use?"

This is perhaps an appropriate framework to talk about the moral uses of power. What would Christ's response be? He said: "Consider the lilies of the field. They neither spin nor toil and yet not even Solomon was clad as well as these." He told us not to worry, to seek the kingdom of God first. I see the energy utilities as a self-perpetuating system. In other words, they create the need, perhaps with advertising and with new technology, and then they fill it. They get us all hooked on it. Do we need it? I've lived on a Navajo reservation as a sheepherder and found life quite pleasant without electricity. I've been in Third World countries, and I didn't feel the lack of things we think we need in the 20th century.

COLLIER   In the rod-burning process, what level of radioactivity are we talking about? Can they be handled by an average person? How can they be disposed of?

SHEAHEN   When we've completed the reprocessing, the waste is cast along with fresh uranium into a new fuel rod. That goes back to the reactor. That takes place behind five feet of glass. The new rods are highly radioactive; they're very nasty. Other elements like cesium, iodine and the rest of the intermediate fragmentary fission products are also highly radioactive, but will decay in something on the order of 10 to 20 years. We know how to store that stuff for a century or so. The really bad long-lived stuff is back in the reactor. The waste that must be stored underground or wherever will last only for a few decades.

COLLIER   Does this pilot technology you've been working on actually break down these long-lived elements?

SHEAHEN   Yes. We take the rod out of the reactor, put it in a lead cask, haul it to the restraining building and bring it up inside a room enclosed in glass. People in an adjacent room use power manipulators to handle it. They put it into acid for chemical separation. After a few chemical steps, the uranium, plutonium, neptunium, americium, all the long-lived stuff, which stays together, winds up in pile A. Pile B, the shorter-lived waste, is sent to long-term storage facilities. We add the fresh uranium to pile A and make a new fuel rod which is now viciously radioactive. We put it back in the reactor. That's how it's done.

SPITTLER   The radioactive decay process is essentially asymptotic; we have a very rapid decay of a very highly radioactive material. Even when we're dealing with the long-lived isotopes, the level of activity after 10 to 20 years is a fraction of what it was when it came out of the reactor. When we talk about long-life storage, aren't we talking about storing something that has nowhere near the level of radioactivity that it did when it came out the reactor? Aren't we talking about something that's already significantly decayed? From a chemist's point of view, we're simply talking about something for which we haven't found uses. There may be potential use for radioactive isotopes. It's just that we haven't found any uses for them yet. What will people do in a 100 years if we simply disposed of these things in an irrecoverable way. We might discover uses for radioactive isotopes that now we call waste, much as in the 19th century they disposed of tar products like charcoal because it was considered to be waste. The technique for separating the radioactive isotopes is pretty straightforward. It's a matter of what we want to spend.

I was surprised at the amount of tonnage (1400 tons) of radioactive materials in this pool. If we compare that, say, with something like coal, we'd get 100 times that much waste in a day from a coal fired power plant. A coal fired power plant uses a 100-car train of coal each day. Then we're talking about waste products from that.

KARLSTROM   It's been said that one gram of plutonium, if properly distributed, could kill every person on the earth. That's how toxic some of this material is.

SHEAHEN   A lot of people go around making up that stuff. Statements like that have many assumptions. They assume that each one who gets that properly distributed one gram fragment, a micronanogram or whatever, gets it exactly in the worst place, like the corridor to the pituitary. Many people, lacking their own scientific skills, grab onto statements and then engage in advocacy to try to make it seem as draconian (sic) as possible. It gets them in front of the TV cameras and on the Geraldo Rivera show. Such statements have no connection to the realistic way a waste product could conceivably be dispersed.

KARLSTROM   I think you're just talking about quantities. I imagine a larger quantity, if it was distributed, could kill every person on the earth. It would be larger than a gram; I don't know what it would be. But this stuff is extremely toxic.

SHEAHEN   You don't want to get close to it. For security and terrorism reasons, the government keeps extremely tight controls over it. I've been in these plants and there are three and four levels of security. I've never gotten near actual plutonium. It's guarded to an incredible extent. A widespread dispersal of a large mass of plutonium is the kind of low probability event that is not on the front burner along side things like children dying of typhoid from contaminated water. We've got big front burner problems in the environment, and something like this is a low probability event which must stay on the back burner.

GRIESBACH   Eric's comment skirts something that's under the surface in this discussion. It has to do with a seemingly prevalent critique from the Left in advanced industrial societies. It's prevalent in Europe and in the U.S., too, to some extent. Essentially, Western societies are increasingly pervaded by a feeling of alienation. Part is attributable to a sense of absence of control. For many, nuclear power is a symbol of this alienation. The critique contends that the symptoms -- self-absorption or instant gratification -- can be attributed to people's sense that they're in control of the conditions of their own life or that they can opt out. This may take us back to Tom's fifth point last night, namely, that technical and scientific education is critical. We may develop into a split society, in part composed of technologically and scientifically sophisticated folks and those who are alienated. That kind of a split society poses real problems for our political institutions and our general ways of running a nation.

CARTER   I've heard that the Navy has been reprocessing its fuel all along, and it's not a novelty to them. I don't know whether that's true or not. Also, is France a successful reprocessor or not?

SHEAHEN   The two answers are: I don't know and I don't know. Those of us in the nuclear reactor research area are not frightened by the word reprocessing. In the early '70s near Buffalo, New York, in a town named West Valley, there was a nuclear waste reprocessing plant. It worked well but lost money and was shut down for economic reasons.

BERTRAM   Was it privately owned?

SHEAHEN   It was owned by a consortium or something like that. It had a heavy government subsidy, but the operating costs were such that they couldn't make a buck, so it went under. I wouldn't be surprised to find reprocessing going on in France and working out fine for different economic reasons.

CARTER   Is the Navy practice a military secret? I have had the understanding that they do it.

SHEAHEN   No, I wouldn't think it's a secret. Good for them, if they do it. Many do reprocessing; it's difficult, but technically achievable and it has been achieved. It's expensive, so we must be motivated by something other than standard American commercial economics to engage in it. French economics may be different. The Navy may have a different attitude. The reprocessing of first generation nuclear fuel rods isn't economical. It's a fundamental concept in the second generation plants where it's economical.

BENNETT   Tom, you used the word "draconian" to describe the conservationists' interpretation of how lethal plutonium can be. I wonder if the word might also be applied to the people who invested in the first generation of nuclear power plants. Many policies were made about the amount of energy which should be produced, only to find out that the process was not as efficient as thought. Now we have these tons and tons of wastes to deal with. I also want to mention the electricity demand forecasts you showed us the other night. One optimistic group said it presumes that we can go for the next 20 years at least with no increase in the demand for electricity. This has got to be optimistic because most people are not going to do away with creature comforts. Holding the line on energy demands can't be achieved without conservation. First and foremost we must stress conservation.

SHEAHEN   Absolutely. I used the word draconian. I regret that you associated that with conservation. I'm a conservationist and I like energy conservationists a lot. I don't agree with the elements of the environmental movement which say we must absolutely slam the lid shut on population. Many of them push for abortion and that sort of thing. I'm not a zero growth person, but I definitely think that energy conservation is the way to go. In the late '60s, Amory Lovins was a voice crying in the wilderness. Most dismissed him as a fool. In 1973, when the first oil crisis came along, a few people realized that he had written some things. Time passed, and by about 1980 the numerical amount of energy used was the amount that Lovins had predicted. The electrical utilities which built what we now sneeringly dismiss as dinosaurs, first generation nuclear plants we don't really need, were embarrassed. They lost revenue and stockholder benefits and so on because of excessive expenditures. Nonetheless, Lovins had been ridiculed. More recently, he has continued to urge a conservation agenda which is -- let me get off the word draconian -- certainly extreme. In my best judgment, we can't reach the new numbers that he has predicted. But I will say two things. One, we have consistently done better in conservation than predicted. Two, given that Lovins was once right, we have to give note to him. While we may argue and disagree on numerical values, the general thrust for conservation is one that is very definitely present.

Do you remember last night's graph on per capita energy? America was about double Western Europe and Japan. There's no reason why we can't live on the energy budget that Western Europe and Japan live on. As the 3 billion in China and the rest of the world come up the ladder, they will want to come up to the Western European/Japan standard. That's going to take a lot of power. A conference like this one would not happen in some of those poorer countries because they don't have electricity. We could have daytime sessions and then go home. There are so many things that we take as run-of-the-mill items it would be extremely difficult to turn back the clock and revert even to a 19th century energy use. The Little-House-on-the-Prairie thing, which to us would be a colossal step backward, would a major step forward for 3 or 4 billion people. That's the gulf that separates us in energy use.

KARLSTROM   It's a question of values. To me that lifestyle is attractive and Las Vegas' wasteful use of energy is obscene. Let me quote Barry Commoner's The Poverty of Power. "In the last ten years, the United States, the most powerful technologically advanced society in human history, has been confronted by a series of ominous seemingly intractable crises -- the threat to environmental survival, the apparent shortage of energy, and the unexpected decline of the economy. These are usually regarded as separate afflictions, each to be solved in its own terms: environmental degradation by pollution controls, the energy crisis by finding new sources of energy and new ways of conserving it, and the economic crisis by manipulating prices, taxes, and interest rates. But each effort to solve one crisis seems to clash with the solution to the others. Pollution control reduces energy supplies. Energy conservation costs jobs. . . . This tangled knot . . . involves complex interactions among the three basic systems the ecosystem, the production system, and the economic system that together with the social and political order govern all human activity. . . . The economic system is dependent on the wealth yielded by the production system, and the production system on the resources provided by the ecosystem. Given these dependencies, the economic system ought to conform to the requirements of the production system and the production system to the requirements of the ecosystem. In actual fact, the relations amongst the three systems are the other way around."

BERTRAM   Tom, we all seem to agree that conservation is the first priority. It's at least conceivable that we could improve our conservation practice in North America, especially in the U.S.A. You're not suggesting that that would be sufficient to raise the "rest of the world" on your graph to a decent standard, are you? Won't it take something else beside conservation to make that kind of energy available for them?

SHEAHEN   Conservation ought to lower our ridiculously high level to the Western European/Japan level. That demands changes in lifestyle such as taking a bus instead of driving our cars everyplace -- that level of thing. But conservation can't help someone who has no energy in the first place. When a hotel in Nairobi gets sufficient electricity for its lights, when a nation is increasing the general level of energy use, then we hope that conservation is included at each step. But they can't conserve what they don't have in the first place. Their basic energy needs for electricity will have to come either from coal or nuclear or it's going to come in very small doses from hydropower, solar or wind. If it comes in very small doses, their national development is held down. They can't grow, they can't purify their unclean drinking water, they can't lower their death rate. Many such events will happen until they achieve that first big step upward to be major users of electricity.

BERTRAM   You're not saying that you see a kind of reciprocity between a shift toward out forbearance and their progress?

SHEAHEN   Yes. As soon as we do something in conservation. If, instead of building another hydropower dam, we insulated houses, money would become available for bankers to lend to Brazil, Uruguay, and so on. It would allow capital to go to work elsewhere in the world. This American conservation would definitely help the less developed countries because it has that effect on the economic ratio.

BERTRAM   Is there economic experience to bear out the optimism that, if that capital were freed up, it would instead be invested in developing countries?

SHEAHEN   We'd have to work hard to make the connection because, at least on a local level, if I had an extra $20, I'd probably buy a ticket to the hockey game. I don't think bankers necessarily take global views. But by and large, it tends to happen. There's constant pressure from the underdeveloped countries to get some kind of income and revenue. The available money has to come from the developed countries who are challenged within themselves to engage in building new power plants. If, because of conservation, we are not building highways, schools and so on, money will be available to the people who need it more.

BERTRAM   Through the medium of investment capitalism, we give incentives to the Third World countries?

SHEAHEN   I think that's fair to say, yes.

COLLIER   In the Third World, the primary energy problem is the amount of net income spent on food. This often is, as I said, 40 percent of their annual salaries. As a part of this whole process of energy conservation and helping raise the level of the Third World countries, we need to help them increase their ability to produce food. More, we have to do this with low-cost inputs of energy and fertilizers. They don't have much money to buy fertilizer. When the cost of petroleum goes up, it drives up the price of fertilizer. In this country we hear a lot about sustainable agriculture. We've heard that term this weekend. That's the new buzz word for lowering our energy, fertilizer and pesticide/herbicide input into the environment. Again, the only practical way to do this is either to change our food sources or to engage in a widespread program of biotechnology to alter plants to produce their own pesticide or herbicide for the specific pests that attack them (we won't have to apply it to the soil) or possibly to produce their own fertilizer. Simply saying we're going to lower energy costs won't solve the problem.

SHEAHEN   You're bringing up a very important point here. My paper was grandiosely titled "The Energy Outlook," as though I was going to present the entire world picture on energy. The word agriculture does not occur in my paper nor does it occur in the minds of most of the people inside the Washington Beltway. It's a blessing that we're holding this conference in St. Louis and people from Monsanto are here, because agriculture is probably the most underrated technology in the world. It uses chemistry, biology, biotechnology. It has tremendous potential. How can we meet all these people's energy needs? We can't build 100 nukes each year for a century. Truly innovative solutions must come in areas like agriculture. That may be the pathway for the solution. We speak romantically about the underdeveloped countries and appropriate technology. It often refers to some clever little water pump or something like that. It works for a little community of 250 people or so. On the large scale, solutions will have to come from some very innovative biotechnologies in agriculture.

COLLIER   In this country we spend 12 percent of our net income on food. This is the lowest of any country in the world and it's one of the reasons why agriculture is not a big topic here. If we go to any other country, we will find them spending somewhere between 27 and 75 percent of their net income on food. We can afford to be choosy.

SHEAHEN   Are there countries that hit 75 percent on food?

COLLIER   In bare-existence economies essentially all the time is spent providing enough food to survive or enough fuel to make a fire.

BERTRAM   Does the relatively low proportion we spend for food correlate with the high proportion we feel we can spend for energy?

COLLIER   It gives us much more expendable income for luxuries. A starving individual has one problem, the rich person many.

GRIESBACH   I'd like to ask more about that conservation option in the U.S. Your chart yesterday had the U.S. per capita expenditure a bit less than twice Japan's and Western Europe's. You suggest that we get as low as Japan and Western Europe. Where do you see the action? It seems as if much of it has to be in the personal transportation context. Is there much room in industrial energy efficiency? I understand that that accounted for a good bit of savings over the last ten years.

SHEAHEN   Yes. First, I would rank private transportation at the top of the list. Second, energy efficiency's one of the most amazing things I've ever seen. I was very active in that field in the late '70s and I did a lot of calculations on how far we could push the paper and steel industries and so on. As weird as it might seem, as long as we're still far above the thermodynamic limit of the chemical event's energy transaction, it looks as if we can still have a potential savings of 25 percent, no matter how much we save. And new things come along. New types of electric motors and so on that we didn't imagine in 1978 are making their way every day into plants. If I tell you today that we have almost exhausted all the energy conservation opportunities, tomorrow some one will invent something that will prove me wrong. Each year, no matter how much we save, it looks as if we can save 25 percent more. We'll get to the thermodynamic limit some day, but we're nowhere near it for any important industrial process in this country.

GRIESBACH   Nevertheless we're going to have to rely on the private transport sector for a great deal of the conservation.

SHEAHEN   Well, buildings, too, of course. Buildings in America are not nearly as energy efficient as they are in many other countries. But yes, transportation is absolutely at the top of the list.

BERTRAM   Does air conditioning have to go?

SHEAHEN   Well, countries like Switzerland where it never gets very hot, don't have air conditioning. These are some of the trade-offs that Americans don't want to make. In the 1940s and '50s, when few had air conditioning, our energy use was probably higher per capita because people were wasting it elsewhere. If I were in Congress and if someone took the gas guzzler to the landfill and then bought a 40 mile per gallon car, I'd give that person a "credit" to have air conditioning. There's a disproportion between American waste in the transportation sector and the relatively small American waste in the building sector. It focuses attention on the fact that we can't ignore mass transportation.

BRUNGS   I can't understand why we've refused to build a high speed rail network this country. From the standpoint of energy conservation, rail is as low a per capita expenditure of energy for transporting people or things as we have. But we spend on jet engines which, though they are more efficient than they were, still consume a lot of fuel. If we had France's rail network -- we're a much bigger country, but it can be done -- we would save an enormous amount of energy. I'd take a train from here to the west coast if I could get there in 20 hours. But we adamantly refuse to go into rail transport. I have written to the last three presidents suggesting that we ought to rebuild our railroads.

More, we're giving up the Panama Canal in a few years. How are we going to move large bulk cargo rapidly if we have to?

SHEAHEN   Right! Neglect of the rails is a great contrast between Europe and us. It's a lesson we could learn from the Europeans.

SHERMAN   People should not get too scared about plutonium being so toxic. It is toxic, but the real danger is its being diverted for bombs. There are probably hundreds of other chemical compounds in this country and around the world, in much larger tonnage than the plutonium, that are just as toxic, if not more so. These are being safeguarded every day in the chemical industry -- arsenic compounds, cyanide and so on. I don't think the toxicity argument holds any water.

We can't compare our transportation to Japan's or Western Europe's because of differences in population density. Where the population is densely located, say along the east coast, we have good rail services. Where it's spread out, we don't. It's not reasonable to think that we can approach the energy efficiency of the Japanese or the Western Europeans, because we don't have the population density to support that kind of transportation. We must go for other conservation methods like more efficient cars. Maybe we ought to put some kind of cap on the weight of cars. If we allow heavy cars on a highway while we're encouraging light cars, we know which one gets the worst of the deal if they collide. If they're all small cars at least each has an equal chance in a collision.

There are many other things we can do in conservation. When I traveled through the Japanese countryside two or three years ago, I was impressed that practically every home in the countryside had a solar installation on the roof. Almost every single house heated water with solar installations. We don't see that in the U.S., but we certainly could do it. There's a lot we can learn from Japan and the Western European countries, but we shouldn't fool ourselves into thinking that we're ever going to have the types of energy density or usage that they have.

CURRIE   We haven't answered the serious question Eric raised. Are we willing to lower our standard of living or must we lower it, for the Third World to improve? If we must, are we willing to do it? There's evidence that we're not, and that creates a real problem. I suspect that most of us aren't aware of how bad things are in the Third World. I don't think we realize that debt repayment to the developed countries is greater than the amount they spend on their own people. We're cutting foreign aid even for little countries like Honduras, one of the poorest countries in the world, certainly in the western hemisphere. We're trying to halve our foreign aid there, and that's only something like $50 million. There's a serious lack of leadership on any kind of foreign aid at present.

Finally, look at the reaction that we have to tax increases in this country. Political leaders are afraid to mention taxes. That's indicative, I think, of the fact that we're really not that interested in helping others. This brings us to the basic question of motivation. How do we encourage a sense of responsibility to creation, to people around the world at the same time we pursue important scientific and technological answers to some of these questions? Eric has touched a basic issue that gnaws at all of us to some degree. But we don't really have the answer to it.

BENNETT   When we think about how much the Third World spends on food, we must consider some important economic issues. The capital already invested there has been misspent; many of these countries are trying desperately to increase their exports to get enough capital just to pay off the existing loans. Brazil is in desperate straits. They're not tearing down the rain forests for profit. They're doing it because they need export crops. Often these are particularly wasteful to the environment by exposing the soil to tropical rain and sun or depletion of the nutrients. We have to take the economics of the Third World seriously before any ecological considerations can be met.

My next point is conservation. We should scrap and recycle the gas guzzlers to save landfill space. As for air conditioning, simple conservation methods are possible in almost every home. Simple things like caulking, weather stripping or using curtains, as well as proper insulation, storm doors and windows could drastically reduce energy use in a home. We don't have to give up air conditioning; we can use it more efficiently. Despite all our conservation methods, we're probably still going to need more energy for the future. So, I want to ask about the future of fusion. I believe Argonne has been doing research on it. Are you familiar with McDonnell-Douglas' research into hot fusion?

SHEAHEN   First, cold fusion is a bust. Hot fusion research continues and they are definitely making progress, but the reality is far away. In my paper I said not to count on it. Most people knowledgeable about the problems of building a large size reactor talk about a 30-50 year time frame. That's the time frame even if a breakthrough occurred tomorrow. The people at the Princeton Research Lab keep claiming that the breakthrough will be tomorrow. But they've been saying that for ten years. The confidence in fusion as a viable energy source is just about zero. We can hope and pray for it, but we can't count on it.

BRUNGS   Didn't somebody break even recently?

SHEAHEN   This is one of the things that bugs me about people putting out press releases. They got a certain achievement using deuterium. If they could have done it with tritium, the number would have equalled breaking even, because tritium is a 30 to 50 times better material. Now they're going to try to do it with tritium. Maybe their press release will come true. Unfortunately, the statement probably was meant to promote renewed funding from Congress. It's tragic. When enough things go wrong, when nothing happens or the promises never come true, it will erode confidence in the future of science in the public mind. The recent statement that used the word "break even" was misleading.

BRUNGS   About two years ago an historian of technology, George Basalla, came out with a book in which he showed that every new energy source has been trumpeted as "Paradise Found." Coal, electricity, oil, nuclear power, fusion -- each has been proclaimed to be the final answer to our power needs. In 1955 von Neumann predicted that by 1980 electricity would be so cheap there would be no sense in metering it. It would be cheaper than air. Von Neumann was certainly nobody's fool, but he was a bit euphoric when he said that. We never learn.

SHEAHEN   Even solar power which everybody agrees, "Hey, it's free, what can possibly go wrong?" has problems. We can make solar cells out of silicon one centimeter at a time. I'm sure it's more than a centimeter by now; it may be as big as a magazine or something, with a new process. But the kind of solar cells we need to provide energy to some of these countries, may require a solar collector the size of Austria. The capital required to carry out the manufacturing process of those solar cells puts us up into the billion dollar range.

BERTRAM   Well, folks, was it Abraham Lincoln who said that the best thing about the future is that it comes only one day at a time? This day is just about over.

GENERAL DISCUSSION

SESSION VI

GRIESBACH   I'd like to present a skeletal overview of the Clean Air Act regulatory structure, partly to address the problem of regulatory overload Bob Brungs mentioned yesterday. He called it the omnipresent, omnipotent hand of regulation. I think we can get a sense of the problem by working through some of the structure of the Clean Air act. Then, by looking at aspects of the Clean Air Act regulatory program, we can open up some of the economic issues that Nancy Kete brought up in her paper.

The first issue is the "hyperlexis" problem -- too much law. My own sense is that much of the complexity simply reflects the technological, economic and political complexity of late 20th century life. Part of it is driven by a cultural phenomenon. I'm not sure what to call it. I tried to get at it a bit in my paper. Maybe we can work it out as we go through this.

Let me approach the problems as they developed. Before 1970 there was little federal air pollution control effort. As in most developments in these areas, the states and municipalities were pretty much allocated the task. The environmental revolution of the late '60s and early '70s led to a federalization of air pollution control, and in '72 we got a federalization of water pollution control. The 1970 Clean Air Act set up a two-pronged approach to air pollution control. One established what are called national ambient air quality standards. There was a primary standard and a secondary standard for what were at the time five and are now six "criterion pollutants" -- particulate matter, sulfur dioxide, carbon monoxide, nitrous oxides, ozone. These standards were established to control hydrocarbons and nitrous oxides. In '82 or so, lead was added.

I have in this outline [See the following page for a summary of these standards.] the existing primary and secondary national ambient air quality standards with respect to these various criterion pollutants. The primary standard is a health based standard. I believe the statutory language specifies that the primary standard is to be set at a level sufficient to protect public health with an adequate measure of safety, whatever that might mean. The secondary standards are what are called voluntary standards relating to non-health based damage -- corrosion of buildings and of paint, visibility and so on. The national ambient air quality standards were set by EPA in the course of a rather elaborate set of regulatory hearings from 1971 through about 1973. The act basically challenged each area and required EPA to promulgate the standards. The

1970 CLEAN AIR ACT

1.) NAAQ (200 + AQCAs)
2.) NSPS (BACT)
3.) New auto emission limits (90% reduction)
4.) SIPS for existing sources (RACT)
5.) NESHAPS

1977 AMENDMENTS

1.) Non-attainment ("dirty") areas
  a. SIPS rollbacks - BACT/TCMs
  b. New/modified major sources
     LAER
       offsets
  c. ERC's/Emissions banking
2.) PSD ("clean") areas
     Class I - pristine
     Class II
     Class III - up to AAQS
  a. Increments

  b.) Visibility
  c.) Pre-construction review for 28 categories of new sources
  d.) Offsets

NAAQ = National Ambient Air Quality
NSPS = New Source Performance Standards
SIPs = State Implementation Plan
PSD = Prevention of Significant Deterioration
NESHAPS = Nat. Emissions Standards for Hazardous Air Pollution
TPY = Tons per Year
ERC = Emission Reduction Credit Program
PPM = Parts per million

1990 AMENDMENT

1. Acid rain - 10 million TPY reduction/SO_2
               2 million TPY reduction/NO_x
   (9 million TPY left distributed as marketable "allowances.")

2. Phase-out of CFCs

3. Autos: 40% < HCs/60% < NO_x

NAAQS

		Primary	Secondary
PM-10	AGM	50 $\mu g/m^3$ *	Same
	24 hr	150 $\mu g/m^3$	
SO_2	AAM	80 $\mu g/m^3$ (0.03 ppm)	1300 $\mu g/m^3$ (3 hr) (0.5 ppm)
	24 hr	365 $\mu g/m^3$ (0.14 ppm)	
CO	8 hr	10 $\mu g/m^3$ (9 ppm)	None
	1 hr	46 $\mu g/m^3$ (35 ppm)	None
NO_2	AAM	100 $\mu g/m^3$ (0.053 ppm)	Same
O_3	Max daily 1 hr Avg.	235 $\mu g/m^3$ (0.12 ppm)	Same
Pb	Max Quart Avg.	1.5 $\mu g/m^3$	

* $\mu g/m^3$ = micrograms per cubic meter

standards, applicable nationwide, are intended to set a national ceiling on these various criterion pollutants.

These pollutants were selected because they're the most widespread, and most easily recognized technically, threats to health risk in the ambient air. As you can see, for each ambient air quality standard for each pollutant, there is an averaging period that varies from the annual geometric mean for a particulate matter 10 (PM-10 in the outline) to the one hour mean for carbon monoxide, with corresponding maximum concentration levels nationwide. Particulate matter-10 is a relatively recent change. The 10 refers to particulate matter smaller than 10 microns. The previous standard had been what was known as the total suspended particulate matter standard which included particulates of varying size. As a result of quite a bit of study in the '70s and early '80s, it was pretty well recognized that particulates over 10 microns were likely to be either kept from coming into the human body or expelled by human respiratory processes. The focus is now on particulate matter of less than 10 microns.

There's uncertainty in the selection of each of these standards. Part of the uncertainty is the meaning of "necessary to protect public health." Who's public health? Is it the health of the average American or that of the average victim of emphysema or the average child asthmatic? The statute doesn't say. There's been a lot of a kind of cost-benefit analysis. The costs are rough, seat-of-the-pants, guesses; the national costs of coming down to these standards are compared to the likely health benefits of targeted groups, with judgment calls all the way through.

BLASCHKE   Are the standards for an industry, city or state?

GRIESBACH   No. The standards are ambient air quality standards. They set a ceiling to the maximum concentration at which any of these criterion pollutants can be present in the ambient air.

BLASCHKE   So in Glenville, Oklahoma we have the same criteria as does St. Louis.

GRIESBACH   These are nationally applicable standards. We've got the same particulate standard in St. Louis and in the Grand Canyon, anywhere. The implementation is very different, but the standards are the same. There is obviously a great deal of difference in implementation.

Where do you monitor them or how often? There are many possibilities and much room for the manipulation of the standard once we've got it. There has been a great deal of litigation and a lot of regulatory activity with respect to all of that. In any case, we could say the heart of the '70 Act was the promulgation of these national standards. The remaining sections of the '70 Act in a sense related to devising mechanisms to realize the accomplishment of the standards. In 1970, many areas of the country, particularly but not exclusively urban areas, were at various levels above the standards. It was an attempt to roll back existing air pollution across the board, especially with respect to the ozone standard.

There were three basic regulatory mechanisms in the '70 Act. One was the development of a program to regulate new big stationary sources -- the New Source Performance Standards (NSPS). EPA was charged with promulgating technologically based regulations, employing the best available control technology. With respect to new power plants of various sorts, new paper mills and aluminum reduction plants, there was, in effect, a regulatory attempt made to force technological improvement in the pollution control reduction area with respect to these six criterion pollutants as well as to any pollutant coming out of one of these major new sources that was regarded as raising health or welfare risks.

The national ambient air quality standards are applicable nationally, but the nation is divided up for purposes of air pollution control regulation into over 200 air quality control regions. That's what these non-attainment ("dirty") areas and "clean" areas refer to in the outline. In some measure the boundaries of these regions are political. Primarily, there's an attempt to identify an airshed with respect to which some comprehensive regulatory effort can be taken to satisfy the standard.

In the early '70s it was thought that you could actually identify airsheds of these various criterion pollutants and control them region by region. It quickly became clear that, while perhaps particulate matter is for the most part a local problem that could be handled by way of air quality control region's CO_2, certainly when we get into SO_2, NO_x and ozone, we're dealing with trans-air-quality-control-region fugitive emissions. In any case, the new source performance standards (NSPS) was one of the approaches adopted to try to satisfy the standards by tightening up on new sources. In addition, one of the notorious programs of the early '70s was the new auto emission control program. The '70 Act anticipated a 90 percent reduction of NO_x, hydrocarbons, and carbon monoxide (CO) by

1975. Given the eight or ten year lead time for new automobile designs, this was understood at the time to be rather optimistic. Indeed it was. It brought in some counterproductive effects. There was a continual game of bluff and counter-bluff between the EPA and the auto manufacturers. Finally, we had the first round of pollution control devices, a technology that greatly reduced the fuel economy and brought about a generation of gas guzzlers in the early and mid '70s. I don't think it was until the late '70s and early '80s that the current generation of catalytic converters really began to approach the 1970 target. In any case, we got new source performance standards for new big stationary sources and also the auto program which attempted to reduce ozone, NO_x and CO primarily.

In addition to these new source technology pollution standards, the Air Act required the states to perform what were called State Implementation Plans or (SIPs) insofar as necessary, if these two national federal approaches wouldn't bring a particular air quality control region down below the national ambient air quality standards. The SIPs were to include controls on existing sources, the development of transportation controls of various sorts, state efforts at changing their own practices like shifting from hydrocarbon-based asphalt to water-based asphalt, and so on. The Air Act indicated that the states, in fashioning their SIPs, were to use reasonably available control technology requirements. This created a problem. How many different categories of emitting activities occur within any relatively complex urban area? The states were notoriously unable to fashion SIPs with any kind of bite or ability to control much.

The '70 Clean Air Act had anticipated that the primary national ambient air quality standards for all of the criterion pollutants but SO_2 and ozone would be reached by 1975 and that the ozone and SO_2 standards would be reached by 1977. We came nowhere near it. I have data showing that 27 states failed to satisfy the ozone standard by 1977. The main progress has been made with respect to particulates. Another problem was a kind of perverse incentive which came out of the 1970 Act. It was an economic incentive to construct new plants or add on to existing plants in clean air, where they could avoid SIP requirements or new source performance standards. Firms which could afford to move production facilities to different parts of states or different parts of the country hunted for clean areas. Part of the move to the Sun Belt (the south and southwest part of the United States) of the '70s can be attributed to the Clear Air Act. In any case, with the 1977 amendments, we have an altogether different approach -- a major division of the country into so-called dirty areas and clean areas. Two altogether new parts of the Clean Air Act set additional standards with respect to air pollution problems in each of these areas.

The so-called non-attainment dirty areas, primarily but not exclusively urban areas, received a whole new series of regulatory requirements. SIPs were now required to employ the best available control technologies, transportation control measures were required. For example, a dirty area had to adopt an annual inspection and maintenance of auto emission controls. In addition, the building of new and significantly modified existing sources in new areas was required to employ what's called the lowest achievable emission rate control technology, "achievable" not understood in terms of economic feasibility so much as technological feasibility. State-of-the-art controls were required on new or significantly modified existing sources in dirty areas. In addition, if one were to build a new or modified source in a dirty area, one had to get what are called pollution offsets. Generally this meant that one had to buy off some existing polluter, put him or her out of business. The emissions that that polluter was emitting were now his or hers. Well, not all of them. One couldn't use them all. The new polluter had to assure the regulatory agency that there was a net reduction in emissions and reasonable progress towards attainment of the primary standards. When General Motors built a new automobile assembly plant in Wentzville, Missouri, it had to buy up and close down a number of dry cleaning shops in order to be allowed to set up a paint line which emits hydrocarbons and so on.

Additionally, the EPA in the mid '70s developed what they called the Emission Reduction Credit Program which permitted existing firms to overcontrol beyond their SIP requirements, if they could, and sell the increments of excess emissions to somebody else who couldn't or wouldn't control down to the required level. So now we have a market setup in emission reduction credits. If one overcontrols beyond the requirements of the SIP, he or she can sell excess emissions to somebody else. The mid '70s and all of the '80s witnessed the development of "banks" for emission reduction credits which could be transferred among polluting activities.

That's much of what Nancy Kete mentioned in her paper. She's talking about the same thing in the acid rain context with SO_2 emissions from power plants. That notion goes back to the Carter Administration's emission reduction credits. Another part of the '77 amendments, the so-called Prevention of Significant Deterioration (PSD), Part C of the Act, is applicable to clean areas. To give you a sense of the complexity here, the country was divided into three classes of air quality regions. The dirty areas were class 3. The pristine, mostly national parks and areas out in the west, were class 1. Class 2 was the areas that had satisfied primary air quality standards but weren't regarded as worthy or politically able to remain pristine. The PSD rules imposed incremental long-term increases on the amount of pollution which could be generated in clean areas or areas that had already satisfied the primary standards. The class 1 increment was very, very minimal. I believe, the increment for SO_2 was 0.002 parts per million on the 24 hour mean. Class 3 areas were permitted to go up to the primary standards. The class 2 areas were more or less held halfway between in an attempt to zone the future with respect to emissions. This was in response to the problem of sources running to clean areas to avoid onerous -- even more onerous now -- pollution control requirements in the dirty areas. In addition, the Act mandated all sorts of preconstruction review requirements for 28 categories listed in it. Additional visibility regulations were imposed as was another offset program much like the emission reduction program.

Let me turn to the 1990 amendments as I understand them. This was approved only yesterday by the Senate and it's on the President's desk. It's thought that he'll sign it. The major change occurs in the area of acid rain. I believe it mandates a 10 million ton per year reduction of SO_2 and a 2 million tons per year reduction of NO_x. I also believe the reductions are required by the year 2000, apparently leaving an allowable SO_2 emission total of 9 million tons per year.

This 9 million tons per year is apparently allocated to existing power plants. They have to control down to that level. For all existing power plants -- I believe it deals with only coal fired power plants nationwide -- there's an allocation of 9 million tons per year. I understand from Nancy Kete's presentation, that that allocation becomes a marketable allowance; this permits individual power plants presumably to close and sell their allowance to another power plant. Or they may overcontrol and sell their allowance. They may use low sulfur coal or overscrub the emissions and sell their allowance. The idea is to get increased pollution control by the exchange of marketable permits among power plants.

Additionally, as I understand the 1990 Act, it phases out CFCs (chloro-fluorocarbons) in response to the ozone layer problem. It initiates another round of attempts to deal with the photochemical smog problem, by demanding a 40 percent reduction in hydrocarbons of motor vehicles, a 60 percent reduction of NO_x by, I think, the year 2003. In addition, the old 50,000 mile catalytic converter warranties are now extended to 100,000 miles. There's also a regulatory requirement to clean up fuels. I think it carries a clean gas blend requirement.

Everything I mentioned in the statute is probably an oversimplification. It sets up a rather elaborate process of regulation. The Clean Air Act through the '77 amendments required 118 pages in the U. S. Code. Also, there are four volumes of regulations in very tiny print implementing that. Furthermore, the regulations really aren't the most extensive regulatory detail. That is found in the EPA's day-to-day announcements of informal regulatory advice and so on. It's quite complex. It's in large part an attempt to respond to what seems to be a common problem, namely, that air emissions insofar as they cause damage to health are fugitive. They move all over the place. Dealing with them is a mess.

ABELL   Thank you, John. You've taken away some of the work that I thought I would have this morning on Nancy Kete's paper. There are still a number of things in it that I'd like to talk about and respond to. There are two problems I'd like to mention: acid rain, which I'll talk about maybe a little later, and the oxidants as illustrated by ozone. Ozone blocks the acetyl nitrates and some of the others. But the main thing the Act looks at is ozone. Even with the 40 percent reduction in terms of hydrocarbons and 60 percent of NO_x, they've got a real problem. I doubt if that can be attained, but we'll see.

In terms of SO_2 compliance, we've made great strides with respect to local air pollution quality standards. But the problem is an aggregate of SO_2 emissions which are taken up by the prevailing wind and moved from one location to another. They all seem to be gathering in the eastern and northeastern United States. That realization is responsible for the 9 or 10 million ton reduction figures for SO_2 emissions. But locally in most areas we're doing a pretty good job. I have some experience with the standards that you've listed on the board. For a number of reasons I got involved in this in the late '70s with some industries in the Alton - Wood River area (in Illinois).

Many industries could generate their own power, so they switched to their own power plants. They were burning coal. They made the decision to convert to natural gas. During some of the harsh winters of the late '70s, they were informed that natural gas might not be available to them. They wanted a variance to burn coal when natural gas would not be available. So we wanted to take a look at the difference in particulate pollution, for example. We chose particles because, while we didn't have that much data on SO_2 and SO_x, we had a lot of data on particles and we wanted to take a look at conditions in the past. Is there a significant difference in particulate pollution from one event to the other on coal burning days versus non-coal burning days?

With the establishment of the air quality control regions, the Alton - Wood River people, utilities as well as some industries, did not want to get into the same region as Granite City. As far as particles are concerned, Granite City is the dirtiest of the dirty in and around metropolitan St. Louis. They were being included and they wanted some ammunition about air pollution mapping and whether or not they were within the air pollution plume in and around the St. Louis area. EPA came up with a study that said they were within the plume because the prevailing wind is out of the south. In the summertime that is correct; the wind is out of the south, although we do get moving weather systems and a great deal of wind variability. In the wintertime, it's certainly not out of the south. It's from the west-northwest. I hate to use the word monsoon and I'm not using it strictly here. But there is a monsoonal tendency in terms of wind fluctuations from season to season within the area. We looked at that for them. I might mention the data network that we had.

The sensing sites for the sulphur oxides were few and far between. It's a very expensive undertaking. It's very sophisticated in terms of the equipment and the interpretation of the results. The particulates are relatively easy to obtain. All we have to do is have a screen and move air through it. We collect the material on the screen over a certain period of time and take that to a laboratory where it's dried out and weighed. In effect, that's what they were doing then. In and around the St. Louis area we had a fairly decent tight network of particulate sensing in the city of St. Louis. We put many of them at or near fire stations where they were a little more secure. In St. Louis County, we had a pretty good data network inside Highway 270. Outside Highway 270, the sites were few and far between. We had a dense data network in the Granite City area. The EPA had them; we had them and the Granite City air pollution control board had sites at three of the firehouses in Granite City. After that, the sites were just scattered here and there. I did manage -- and this was kind of a coup in those days, although I had to promise not to do certain things with it -- to get material from three Illinois Power sensing

238

sites in the Alton - Wood River area. Strangely enough in an area like Sauget (heavy chemical industry site), we had no regular sensing sites.

We plotted numbers on maps. We were using geometric means for the particle counts. It's still quite an undertaking. It's inexpensive to do in terms of collecting something in a screen, but it has to go to the laboratory. It was decided to sense this only every sixth day. During a year there are so many Sundays, Mondays, and so on. We could see differences between the weekend when the industry is geared down and the weekdays when we expected more particulate pollution. We randomized the first day that we would do it, using a table of random numbers.

We came up with three dirty spots. Granite City was the dirtiest. The mouth of the river Des Peres in South St. Louis County and South St. Louis City was another very dirty operation. The Alton - Wood River area was another maximum. The maximum in Granite City was very large. I couldn't determine a separate maximum for the city of St. Louis nor could I do it for Metro East. In Metro East we didn't have the data available. The city of St. Louis blended in with the maximum in Granite City and the maximum in South St. Louis, so we had a lot of air pollution in between but we had the two maximum areas up on either side. I would have liked to map the SO_x at the time. The data simply was not available.

Even taking something that's relatively cheap technologically, relatively easy to do, we still do not have a complete sensing network. The particulates are interesting. We were pointing to industry. In addition to the fact that it is a particle, there may be some problems as to exactly what that particle comprises (is it soot or is it something else?). We can go to White Pine County, Wyoming and get a particle count. There are natural particles out there. We're not divorcing the natural from the so-called man-made at this time. We get a particle count no matter where we go. I selected White Pine County, Wyoming, because, in all the inventory I've seen, it has the smallest number in that particular area.

Many things put particles into the atmosphere. A large road construction project in an area for some years is a very dirty operation, but we don't have anything coming out of a stack. A grain handling business is a very dirty operation as well. In terms of total particle inventories, going back to the data I used in the '70s, 1/3 of the particles we were counting in the atmosphere came either from forest fires or from agricultural burning. Of the forest fires, 1/3 were, we felt, man-caused and 2/3 were natural, almost all of the latter, of course, by lightning. So there's a lot out there in the atmosphere, some of which we have to consider to be quite natural. As far as the SO_x are concerned, no matter what some of the activists demand, we're not going to get down to SO_2 counts of zero. There always will be some. The approach that's been taken in the past has been realistic in terms of local problems.

There's a synergistic effect between particles and SO_2. By that I mean that, if we take certain bad health effects from particles and from SO_2 and add the two effects together, we get a certain number. If we get the particles and the SO_2 in the area simultaneously, which we usually do, the combined effect is worse than the additive effect of each taken separately. For example, consider what happened in London in 1952 where weather was an accessory before, during, and after the fact as far as mass homicide was concerned. We're talking conservatively of 4,000 to 5,000 deaths that we attribute to an air pollution incident between December 1 and 5, 1952 when a high pressure cell came over the Thames valley. It just sat there; there was relatively little air movement so there was no ventilation sideways. This threw an inversion over top of the Thames valley so the material couldn't be moved off. Initially there was a fog, whitish in color; it was relatively clean. Because of the burning of fossil fuels, this gradually built up to lethal levels; predominantly people in poor health were the ones who were adversely affected. On December 5th a cold front moved through the area and cleaned up the atmosphere. There was no stagnation after that, but people continued to die because some evidently had had their respiratory apparatus so adversely affected that bacteria and other things could take over. We're looking at statistics about how many should have died during the period and how many did. Was there an organism around that might explain the excess deaths? I have a vague memory that the number of smoke particles was about 750 $\mu g/m^3$. Smoke and sulfur became an extremely deadly combination.

I'd just like to bring up a couple of other things here. By the way, CO_2 is not a local problem; it's regional and worldwide as is SO_2 and NO_x. For automobiles, the first steps there were the California regulations. I suspect that the early federal regulations were taken from the work done in California. In California in the late '60s they first attempted to increase the air/fuel ratio in the internal combustion engine motor vehicles. In doing that, they ameliorated the CO and hydrocarbon problems. They got good reductions. They were looking at three curves there: The CO_2 was rising year to year; the NO_x was rising; the hydrocarbons were rising. They went in with the first controls and the big one, I think, in the first controls was the reduction of the increase of air/fuel ratio. Immediately they departed from the projected curve and started to see a reduction in CO and in the hydrocarbons. The NO_x took off, because in increasing the air/fuel ratio, we bring more air through the engine. This increases the NO_x. As far as the internal combustion engine is concerned -- I'm way out of my field now -- but I would think they're doing things like retarding spark and reducing compression ratios and things like that to help to counteract that increase. They've had some success with that.

It's only natural that these attempts came out of California, particularly the Los Angeles area. They have four things going on there that gave them a classic photochemical smog problem. They had a lot of cars. They had a terrain effect where effectively people live in a bowl. They had a lot of sun. It's one of the things they used to brag about. In photochemical smog, the energy in the ultraviolet wave lengths from the sun gives a whole plethora of substances that never saw the inside of the internal combustion engine. Take ozone! If we try to measure ozone in automobile exhaust, we don't find any. Ozone is one substance along with peroxy acetyl nitrate and the aldehydes that would emerge in the air; they peaked at different times. This emphasizes that we're looking at a kind of chain reaction with the photochemical models.

Of course, they also had a climate problem, a very stable atmosphere and a cold current off the coast. In the meteorological dynamics, they have a Pacific high pressure cell, as exists in the Atlantic and Indian Oceans. The inversion starts on the average of about 25,000 feet over Hawaii and then kind of slopes down toward the California coast. Oftentimes it intersects the coastal ranges. This is in addition to "living in a bowl." For these reasons, many of these regulations came from California.

I think ozone, photochemical smogs and oxidants, will be the real problem in the future. We're on top of the SO_2 problem locally. But the aggregate problem in terms of contributing to acid rain, which perhaps I could talk to for just a few minutes later on in the conference, is the problem we're facing right now. We've made progress here with respect to health. Again, we can get to the argument of the chronic sufferer. I'm glad John Griesbach mentioned, for example, the asthmatics. What is going to be in some cases critically harmful to them, would not bother people who don't have this kind of a problem. I think there has been a lot of progress, but there's still a lot to be done in the future. The entire technology had to be developed for this. They knew they had a problem in California, but they didn't know what it was. Initially they had to develop the technology to identify the problem and go from there. They made some bad decisions. But the field was in its infancy then; now it's becoming more sophisticated. I think the problems in the future will be acid rain because -- that's an aggregate problem since air pollution knows no political boundaries -- and the oxidants. Quite frankly, I don't see a solution for the oxidants right now with the internal combustion engine.

BERTRAM   Ben, you referred to the study in which you were engaged in this region. For whom was that done?

ABELL   It was done for the Olin Corporation. They didn't want to go into a region that included Granite City. They felt that they would have no incentive to improve and it would mean zero economic growth for that area. They felt the situation was so dirty there that, if they were included in that region, they would be at a standstill. That's why they wanted out of that region.

SHERMAN   What does NESHAPS stand for?

GRIESBACH   It stands for National Emissions Standards for Hazardous Air Pollution Sources. In the 1970 Act, section 112 dealt with hazardous emissions. Almost nothing was done. By 1977 the EPA had listed only four hazardous air pollutants: asbestos, beryllium, mercury and vinyl chloride. I believe a good part of the '90 Act mandates additional hazardous pollutant controls.

SHERMAN   I also want to ask Ben whether those monitors are still in place. Is that why we get an ozone alert every now and then?

ABELL   Yes, they're still in place. The ozone alert is a difficult task and the monitors are few and far between. Illinois Metro East has one or two. In Missouri we have them in and around the metropolitan areas. The particle count is still available and published by Illinois and Missouri. It is published late in the year for the preceding year and one could map the pollutants. Where there's a relatively dense network in the St. Louis area we're continuing to see an improvement in particle pollutants. The three trouble spots, however, still stand out.

SHERMAN   Going by recent news from California, they've issued some statutes that in the future the automobile industry has to make available there cars that run on electricity and propane. Do you have more information on that?

GRIESBACH   That is a part of Big Green, I think. [ed: Big Green is the name given to a massive environmental initiative on the California ballot. It was soundly defeated in the Nov., 1990 election.]

SHERMAN   I don't know how they would make people buy these cars, but they're supposed to be available anyway.

GRIESBACH   I'm not sure if it's a requirement on the budget of Los Angeles or if it's some kind of mandate to the Los Angeles air quality control region to impose it somehow. I don't know how it would be done. It may be simply a matter of funding.

SHERMAN   It may have to do with large fleets such as police, city employees and so on. It's a step in the right direction as far as air pollution is concerned.

GRIESBACH   I should mention that all the way through it the Clean Air Act includes as many subsidy-based or subsidy-type targeted funding areas as it has regulatory requirements. For every one of these regulatory areas, there's a fair amount of funding for various research and development (R&D) efforts or for attempts to develop new models.

BERTRAM   I happened to be next to a man on the plane the other day who was associated with the national gas industry. He said that there are fleets, United Parcel Service for one, in certain west coast cities where the trucks are converting to natural gas. This surprised me.

CARTER   There's been a lot of propaganda on the subject of natural gas powered vehicles. It's a promotional effort from the natural-gas-powered motor vehicles industry.

GRIESBACH   I have a couple of questions for Ben. One, are there carbon monoxide hot spots? I've heard that's it's a monitoring problem at the local level with CO.

ABELL   Yes, there are hot spots for CO. The motor vehicle and the way the traffic moves are implicated. It's in the mode of operation of a motor vehicle: idle, cruise, acceleration, deceleration modes. Before controls, only 60 percent of the hydrocarbons were coming from the engine and through the tail pipe. We were getting about 20 percent from the crank case blow by, and about 20 percent from the carburetor/fuel tank system. Even when one turns the vehicle off and it's hot we get some hydrocarbon emission. It's what we call hot soak/cold soak. In cold soak you're not getting much but in hot soak you are.

We have to deal with several factors in relation to CO from automobiles. First is the internal combustion engine itself. Then we have to deal with the weather, for example, little air movement or possibly a thermal inversion. We have to include traffic density (rush hour, for example), the speed of the traffic and other factors to deal with local hot spots for CO. Obviously, there will be industries where we get CO emission. Generally, I see it -- I'm not prepared to say what percent -- as a motor vehicle problem as much as anything.

GRIESBACH   Two more technical questions. You were talking about particulate matter. Does that break down along the size of the particle? I understand that the rationale for the 10 micron limit was that, for one thing, natural particulate matter is ordinarily, almost always, larger than 10 microns. The other question has to do with the photo-chemical smog problem. I understand that there's some disagreement, controversy or uncertainty as to whether or not hydrocarbons or nitrous oxides are kind of a limiting condition and whether that limiting condition varies over time. One might be the limiter now, another might be the limiter next week, the sun might be a limiter at a third point and so on. If that's the case, it seems to suggest all sorts of problems in trying to regulate particular emissions on any kind of national uniform basis.

ABELL   The answer to the first question is yes. With respect to your second question, you'd be better off talking to a chemist or an atmospheric chemist, if there is such an animal. I don't think there is such a profession academically, but there are atmospheric scientists who work into that area and chemists who work into that part of atmospheric science. I would blame the sun as much as anything. The sun is a controlling factor here. When the sun is at a peak in terms of zenith and length of day, there is a smog problem in Los Angeles. The episodic problems occur mainly in late winter and fall when they don't have the cloudiness that they had earlier. The problem becomes critical during the day. The morning rush hour is worse than the evening rush hour because the sun can work on the things that have been emitted into the atmosphere.

The key to that cycle, whichever photochemical model you accept -- there isn't *a* photochemical model (there are many) -- seems to be nitrogen oxide, nitric oxide, ozone, nitrogen dioxide cycle that goes on. Certainly both NO_x and hydrocarbons are necessary. In a study in New York and in things we've seen in and around St. Louis, we're getting some ozone levels at the wrong time and in the wrong place. There may be some sort of a natural source that we don't understand yet. It's still incompletely understood. But the sun, NO_x and hydrocarbons all appear to be the key. We can also implicate SO_x, too. We say we're not talking about CO but in part of the chain it also comes into play. It's terribly complex.

BERTRAM   Would this be a good place to mention acid rain?

ABELL   I'll mention just a few points. I'm not going to redo Nancy Kete's paper. We've already treated the regulatory part of it. It is interesting to see what they're proposing in terms of the trade off that will take place in the lower 48 states. Acid rain is a problem that knows no political boundaries. The Canadians are angry with us. It's a problem in Scandinavia and much of Europe as well. For years people blamed SO_x emissions. It does contribute. Once we get the SO_x into the atmosphere and we get water -- liquid water whether it be in the form of cloud, fog or precipitation -- we'll get a reaction which ends up with sulfurous and/or sulfuric acid. So we're talking about an acid mist.

In the local area there is an industrial plant which had several accidents in which SO_3 was released into the atmosphere. The St. Louis County authorities were not notified. It was a problem and there was an investigation and they felt that the problem source was a certain plant that is no longer in business. In a local area if there's fog when that happens, you can literally live on the wrong side of the tracks, downwind from the emission source. If we have frequent fog, we'll have acid mist in the area. People breathe this. As a case of chronic air pollution, it's more of a morning and a late nighttime problem than a daytime problem, because that's when we tend to get more fog, particularly radiation fogs if there's a river nearby. Certainly SO_x is a major contributor to the acid cloud and the acid rain.

An individual addressing our local chapter of the American Meteorological Society a couple of years ago talked about the mountains in the east. He talked specifically about Mount Mitchell, the highest point in the eastern United States. They were investigating the acid rain problem there and noticed they had a problem up and down the mountain. The problem, however, seemed to be concentrated on the mountaintop; Mt. Mitchell was showing a tendency to become "Old Baldy," so to speak. They wondered why, because one could argue for heavier rain somewhere on the slopes as opposed to the top. But the fact is the mountaintop was frequently enshrouded in cloud. As a result, there was a longer exposure to the acid there.

They were looking at SO_2 and at the burning of coal and of fossil fuels in general. Within the last five or six years, they've started to point the finger also at NO_x because we can get nitric and nitrous acid in the atmosphere in the same way. I seemed to read in Nancy's paper that the government is looking at industry in terms of trading off units in this new Act. If I read it correctly, it seems as if people can trade off not only from region to region but they can trade off SO_x units for NO_x units. That's an interesting concept. But, yes, NO_x is involved.

Natural gas, for example, with its relatively high temperature burning, would perhaps create more of a NO_x and a relatively smaller SO_x problem or CO problem, if any. But I don't think that they have looked at the role of the motor vehicle. Even with the controls they're trying to mandate, we are still going to have an emission of nitrogen oxide and, again, the nitrous and nitric acid. I'm not prepared to say what percent nor do I think anyone is. In terms of SO_2 and NO_x and its implication in acid rain, we may be talking as much as 50/50. It can be argued either way. But of the 50 percent from the NO_x, a large segment of that -- I'm not prepared to say it's more than half -- would come from motor vehicles. A lot of the blame still goes in that direction as well as to some of the so-called large polluters. It's a problem which we haven't identified completely. There's a lot of work to be done.

It's the same old story that I mentioned earlier, namely, the rain areas in Centerville and Edwardsville, Illinois. We have definite rainfall maxima there, and we know it's downwind from the St. Louis air pollution plume on the average. It varies from day to day. We suspect some of it is thermodynamic, but we can't factor that in and come up with a number. More than 50 percent of the rainfall maxima could be the result of the city's heat plume in the area as opposed to air pollution plume, interestingly enough. The automobile and its contribution to NO_x and its contribution to the acid rain problem is simply unknown at this time. We can play around with numbers but there's nothing we can check.

SCHNEIDER   Just one note on the nitrogen oxide problem. One of the tricky things in air-supported combustion is the chemical behavior of a nitrogen as it burns in the oxygen. It's a very temperature sensitive chemical reaction. Back in the '70s, when the initial drive to examine auto emissions and also stationary sources was on, there were efforts to try to balance out competing effects. As one moves to higher temperatures -- I believe about 1500°C is the cutoff point -- one gets a more thorough burning of the fuel. Therefore, there is less emission of the excess hydrocarbons. One also gets a greater conversion of CO to CO_2. One, however, starts to produce much greater quantities of the NO_x. As one lowers the temperature, one cuts off the production of the NO_x, but the hydrocarbon and the CO problems are maximized again. So it's a trade off in which there's no absolutely perfect end point.

ABELL   What you say is true. But looking at it still geometrically, I would think there would be a relatively small air pollutant problem at temperatures less than 1500, say. Indeed there is. Because of the insufficient time of the fuel in the combuster, it's kinetically limited, and we've got a real mess on our hands. We've got impurities in there which could at any time interfere with the reaction, that could stop the reaction. We'll never achieve that perfect end point.

GENERAL DISCUSSION

SESSION VII

BERTRAM   I want to ask our essayists about a growing impression I have. Contrary to the Blaschkean angst that I share -- the sense of being overwhelmed by the unmanageability of the problem -- I've caught a more optimistic tone since last evening. I submit that to the essayists for a critical reflection. Am I to understand, as bleak as things may be, that in many respects there is detectible progress and improvement in conservation, in our management of nuclear energy, in our regulation of air pollution? Am I being too sanguine?

SHEAHEN   I think there's progress in all three of them and a few more besides. It is very difficult to see it on a day-to-day basis. Look at the outline of the Clean Air Acts of 1970 and 1990 on the board. A comparison shows progress. What was sneered at in 1970 and backed off from in '77 as unattainable, is in 1990 deemed real enough to try for. Conservation of energy in transportation, homes, industry, and so forth has done wonders in America . We've saved about 40 percent over what the budget would be if we had believed those in 1973 who said it couldn't be done. There is that much progress there. In the nuclear area we've certainly made progress.

One of the themes that I heard a couple times this weekend, which I believe originated with Bob Collier, has to do with the idea that people are risk averse. In making decisions they flee from risk when they perceive that they don't have control over a situation. They are willing to accept risk whenever they perceive that they are in control. People who get in cars and drive are taking a much greater risk than those who take an airplane, but the perception is: "I have control, I'm behind the wheel. On an airplane, somebody else has control." I think a lot of our technologies for the environment, especially in energy, are dominated by such thinking. As I said on Friday night, the role of scientists now is to educate the public, and bring a degree of education to the public about the difference between real risk and perceived risk. First, we would make a contribution to their sense of well being. Secondly, we would have advanced the possibility of many of these better technologies becoming accepted, put into practice, and ultimately bettering the environment.

BENNETT   I would agree with an overall feeling of optimism about environmental and industrial developments. The reason is public involvement. In the late '60s and '70s, although the public really wanted someone to do something about it, the problems weren't understood as well. Government agencies institutionalized the good ideas but did not put enough time, thought or effort into implementing them. John Griesbach exemplified that in mentioning the mandate to reduce some of the air pollutants from auto emissions in a time frame that didn't allow for the design of that model year car. A mistake like that becomes part of the problem. There's been more cooperation between the EPA, other agencies and industry since then. There's been enough regulation to impose a responsibility on the public to educate itself to let industry know what it wants and to inform the government as to how it wants it done.

I see a lot of confusion on the part of the industrialists with regard to sanctions. The government could do more to explain the different layers of regulation and how it applies to each facility. John Griesbach mentioned that there should be research into how we can reduce all the wastes, especially hazardous wastes. There are ways to do it now. Many states coordinate that information and teach industries how to be more efficient and less polluting. We're just getting started in Missouri. It's important to remember that it will take a public realization of its role in a complex society. Changes in lifestyle as well as communication with government and industry are going to be with us during the '90s.

LEGUEY-FEILLEUX   We have an opportunity in the very popularity of environmental protection and the danger of environmental degradation perceived by the public and the media. Some danger is misperceived, but we can say that the greater concern is a plus. This concern is sometimes based on misperceptions of what must be done. I agree that there is a need for scientists to be more involved in educating the public. Public opinion is fickle and in six months we might have another issue that captivates it. Then ecological concern will disappear until there is another crisis. We must take advantage of the present popularity of the issue. Politicians respond, especially in a democracy, to what the public wants. There's an opportunity to do things that the country might not have been willing to regulate and control before.

We must concentrate on the global situation. People are more concerned about what happens within their communities, less concerned about what may threaten them from overseas. Because we won't accept toxic wastes in our neighborhoods, we export them. There are enough irresponsible governments in the Third World prepared to take the bucks and import the wastes. Ultimately that will come back to haunt us. We can't push our wastes under the planetary rug. Eventually they will come back as poisoned food or a poisoned environment. We have an interest in doing what UNEP has pushed. There's already one treaty in place about trade in toxic wastes, but it has to be reinforced. We must use our international frameworks to alert that part of global opinion that is still not aboard.

The 1992 U. N. conference may be a disaster, if we don't have better relations with the Third World. The Third World is conscious of our concerns, which are not theirs. They will respond to our concerns if we respond to theirs, but we have a knack for not doing that. Our political treatment of the Third World is shameful. We talk about our Christian duties while we ignore their needs. We let the hungry go hungry, the naked stay naked so long as they're not of our color or they are far away. We have money but we feel poor because of our tremendous deficit caused by misplaced spending. Our priorities are dreadful. We help the Third World only when we perceive that that helps us. We buy votes and governments. That helps them to some extent since it's still money channeled to some development. When it comes to real development, we won't help unless there's a political payoff for us.

We leave the U.N. in dire straits financially because we're not in control of it. Neither we, the Russians, the Chinese nor even the Third World is in charge of the U.N. Not being in control is not bad because the collectivity can do good development work with money. We have money to give but we must rearrange our priorities. This is worth doing on its own merits. Helping the poor, even if the poor mismanage their affairs, which they sometimes do, is part of our Christianity. The same goes for development. We have to accept waste. I'm not talking about our waste. That we shouldn't accept because we can control it. When it comes to the Third World, if we try to over-control, we won't achieve development. We have to be prepared to tolerate some local tin-horn dictators filling their pockets, as long as a sufficiently large proportion of aid gets into true development. The best development is done collectively by international agencies that don't worry about the flag following the money. That's spending for its own merit, for the sake of achieving development.

Another dimension is interacting with the Third World to support environmental programs. They don't realize they need us and they're not eager to work for environmental programs. We must be ready to support their causes so that they will support ours. In international politics, as in domestic politics, the maxim "you scratch my back and I'll scratch yours" still maintains. It's good politics to be aware of their needs so they're aware of ours. Churchill said that there are no permanent friends in politics, only permanent interests. I'm neither advocating nor supporting that pragmatism, only acknowledging its reality. In the long run, we're better served when our politicians are pragmatists rather than ideologues.

In summary, let's not forget the Third World aspect. We must not forget their non-environmental concerns, in order to mobilize them to help us with our international program for environmental protection.

BERTRAM   I can't refrain from the observation, Jean Robert, that your's might well be the first speech this weekend which combined within one speech two apparently disparate accents. On the one hand, you appeal to us to be concerned about those who are deprived or ignored, because we're Christians who should be moved to sacrifice and care for the others. On the other hand, you mentioned the apparently opposite accent of "you scratch my back and I'll scratch yours." If you'll permit the observation, we haven't addressed the question of how Christians may be both without contradiction -- neither idealists nor cynics -- or how they may in principle be both for the sake of the world. It's an old Christian theological ethics problem, but there are traditional Christian theological solutions to that which don't say "yes, but." Ideally we espouse Christian ideals of sacrifice and giving help to the neglected. But when we come down to earth we have to be "practical." On the contrary, we see both as mandates from the same creator. I won't go into theological suggestions for modelling these two. The two accents have emerged in separate speeches; you put them side by side with commas in between.

LEGUEY-FEILLEUX   Yes, with perhaps one qualification! It's easy in the name of altruism to resort to means contrary to our principles. I'm not advocating that. I'm only saying that, if we want to achieve our moral goals, we must do it within a realistic context. Perhaps idealists are ineffective in their pursuits because they refuse to come to grips with implementation. Many people reject politics because there is corruption in politics. Not all politics is corrupt; there is much of the purely pragmatic in politics, it's true. But we don't corrupt our values by political interaction. Nonetheless, let's not do evil for the sake of doing good. That's the limit in the pragmatism I'm talking about.

BERTRAM   You're well aware that the "idealists" also have a point in saying that the "realists" tend to water down the actual corruption in which politics and even realistic politicians engage. It's a known ethical argument to say, "Well, if we want to get anywhere, we must do this"; on the contrary can we say, "If we want to get anywhere, we must sin"? And the idealists raised the question, "How can you do that with a good conscience?" I suggest that's the question with which the realist must come to terms rather than, "My hands are clean because I had no choice."

KARLSTROM   Bob Brungs said yesterday he wanted Christianity to be an urban religion. I have a slightly different point of view. I claim that wilderness is a fundamental part of the Judaeo-Christian tradition. Christ spent 40 days in the wilderness where he presumably got his spiritual inspiration. John the Baptist, Moses, Elijah and all the desert Fathers went to the mountain, it would seem, to draw their inspiration from wilderness. If we approach a time in this country or in the world where wilderness is not maintained as an intrinsic value, in and of itself, we risk losing a tremendous source of spiritual as well as biological power. The wilderness is the biological breeding ground of our species and of the whole evolutionary process. We can tamper with it, we can change it as we are doing, but my own feeling, scientifically and spiritually, we had better leave some of it alone. There's not much left now that hasn't been tampered with. I'm one of those radical environmentalists who wants to draw a ring around the places that we haven't affected and try to get some sort of legislative means of preserving them.

In this country and in all the western countries, we view the bottom line as money. That's our final reality. The real bottom line is nature, the great economy that Wendell Berry discusses in his book. We ignore it at our peril. Berry makes the point that eventually the ideal should be to conform our small economy to that of the great economy. He's got an essay called, "Two Economies," and I'll read you just the first part:

"Some time ago in a conversation with Wes Jackson in which we were laboring to define the causes of modern ruination of farm land, we finally got around to the money economy. I said that an economy based on energy would be more benign because it would be more comprehensive. Wes would not agree. 'An energy economy still wouldn't be comprehensive enough,' he said. 'Well,' I said then, 'what kind of economy would be comprehensive enough?' He hesitated a moment and then grinning said, 'The kingdom of God.'"

He's talking about an attempt to merge the natural world and God's kingdom and then to try to conform our behavior to that. Wes Jackson is head of the Land Institute in Kansas. He was head of the department of genetics at, I believe, California State University in Sacramento. He retired and bought land in Kansas where he's pioneering genetic strains which will allow people to grow crops in the prairie throughout the year. These are plants for which one will not have to actually turn the soil and, therefore, expose it to erosion. He's trying to develop an agriculture which will not destroy the environment; it's an attempt to conform the human economy to the great economy. This is where I see hope.

I see hope in the sense that we are part of Gaia, part of the natural evolution of things and not in as much control as we like to think we are. I'll read a little more of Wendell Berry here:

"The key to such a change of mind is realization that the first and final order of creation is not such an order as men can impose on it, but an order in the creation itself by which its various parts and processes sustain each other and which is only to some extent understandable." We scientists have come to the realization that nature is not totally understandable. As we get more answers, the questions multiply. As we get closer to an understanding of how things work, we seem to get farther from an understanding of the total mystery as if there is some asymptotic relationship in which reality is there but we can never quite reach it. Certainly that's the way it is in my field. My optimism is based on the fact that God is ultimately in control. To the extent that we can, consciously or unconsciously, conform our minds and behavior to his requirements, there is hope. Since this is a Christian scientific organization, conformity to him should be our focus. This is the question Charles Ford asked: how can we as scientists conform our work to his demand?

I recently ran across a book entitled <u>Rich</u> <u>Christians</u> <u>in</u> <u>an</u> <u>Age</u> <u>of</u> <u>Hunger</u>, which touches on some of the topics Jean Robert raised. Our economic system allows tremendous disparity of wealth. He calls it structural evil in which we all participate simply by being Americans. It's not our choice. We have so many resources and the Third World has so few. At what point do we take Christ's words seriously and take the shirt off our backs and give it to the poor. The concern for the poor, a continuous theme in the Bible, must become a fundamental concern of our society if we're going to remove this dangerous disparity in the distribution of wealth.

Finally, "resource" is a poor word because it suggests that the world was put here for our consumption. I take Wendell Berry's view again that the earth is the Lord's and we are his stewards. We have a conditional tenancy. We may stay here as long as we take care of the property. Some livelihoods in that sense are right and some are wrong. This should give us a measure to judge what is and what is not appropriate. Berry makes the point that means become ends. We can't create a good product if, in the process, we repudiate the values we say we believe in. The values we evince in our lives will irrevocably affect the goal.

ABELL   I don't know how to follow that. I like a moderate view. I suggest that we can use the earth judiciously, not use it up, not destroy it. I don't see putting a fence around it and saying, "Let it be." It's there for our use, as long as we try to ensure that there is something left for future generations to use. I see an overreaction on the part of some environmentalists, almost deifying the environment. That worries me. But, let me talk a little bit more about air pollution and emissions. I feel good about the way things are going. I think the goals -- those goals John Griesbach put the on board -- are definitely attainable. I even thought so in the '60s and '70s. I didn't see how we could attain them, but I was confident they were attainable. I'm speaking now about the U.S. and Western Europe. We see the industrial mess we have in Eastern Europe. We see the developing Third World and what they must do to try to bring themselves up. I have some pessimism about the latter two. That's where we can heed Jean Robert. They need help and with some sacrifice we have the capacity to help. I think that's where the future lies.

EVERETT   I'll just take up the last statement. They do need help, but giving help requires a willing recipient. The whole process of feeding money into the Third World has been going on since right after World War II. Foreign aid has been around for a long time now. Particularly during the years after the first oil crisis, the Arab nations had an enormous amount of capital to invest. Some of that went into Brazil. Brazil took out enormous development loans and developed like mad. Where did the money go? The figure given is that 1/3 went into corruption and never saw the process of development. It went into Swiss bank accounts or their equivalents. The rest went largely into Pharaonic projects like the biggest hydroelectric dam ever made, into atomic reactors and power stations which now run at about 1/2 capacity because they don't need all that energy. These projects don't contribute to the retirement of the debt. In the meantime, the country can't pay back that debt because of interest payments. The people are hungrier than ever.

If we feed more capital into that country or similar country, we will only repeat the same process, because we still deal with the powerful, never with the hungry masses. Nor do the hungry masses know a way out.

They're almost totally uneducated. I don't see how we can help that. I don't think asking ourselves to make sacrifices to repeat that process is justified. Besides, we don't have the money. This country is broke. We may not like it, but we're broke.

T QUINN   First, I want to thank everyone. I have not been a member of ITEST very long, only about three and a half years, and this is my first exposure to a workshop. I've found it extremely educational. In the business of educating the public, ITEST has contributed significantly to my education. I'm sure we're aware that other groups, sitting in other rooms discussing other subject matter, would talk about the same litany of frustrations we've expressed here -- the frustration of the lack of education of the public and the inaccuracy of the media.

As I mentioned earlier, my game is aviation. I've had some association with it in both the national and the foreign military sales arenas. There is no question about it, we're heading toward greater interdependence. I agree with Jean Robert that there are times when we have to accept things which under other circumstances we might not want to participate in. But we must never exceed our ethical and moral limits. I do a little teaching from time to time about the military/industrial complex and the procurement of military hardware. I'm sure that my fifth grade civics teacher has been spinning in her grave these last ten days watching the exercise that has been going on on Capitol Hill [ed: the annual budget exercise]. That's not the way government is supposed to function. But it is a reality that occurs periodically. When the media is criticized, its usual defense is that it just reflects our culture. I argue that the media influences the formulation of that culture. That being the case, when we speak about education of the public, we have to include the media. I was very pleased to hear Tom Sheahen's comment about education of the public. It's a principal requirement to get beyond where we are now.

In my particular area, I tell my students that they should learn the process and use it to their advantage rather than fight it. Unfortunately, too little of the public knows the process that controls their daily lives.

KANE   I'd like to comment on being able to reach usable conclusions in science and in not viewing the world as a black box [ed: *black box* refers to a system where we know the input and the output but have no idea of the intervening process]. The understanding of the DNA molecule is a major achievement. While many areas remain unknown, uncovering them one by one can be done. A lot of the problems we currently face can be approached by peeling back the layers of the onion and gradually beginning to understand the process using science. If we begin to look at the overall problems of the world, we might become frustrated. How are we ever going to do all that? But, if we take them piece by piece -- I'm thinking mainly of agriculture as a present necessity for the Third World independence through sustainable agriculture -- both our technology and industry has a lot to offer. I can give you some perceptions of where I see the industry going. My remarks do not represent Monsanto. They represent only Jim Kane.

I think that the industry is moving into an environmental consciousness that didn't exist before. Companies in the '90s will begin taking public relations advantage of the fact that they are making products in an environmentally sensitive way and that they're attuned to the consumer more than they have been in the past. As has been frequently mentioned, each of us must take the opportunity to speak about the scientific process to those who don't know about science or don't have science backgrounds. Because it's a continually evolving process, absolutes are few and far between. While we understand certain segments, putting them into the overall picture can sometimes change our understanding. The more we do, the more we learn. It's difficult for an industrial scientist to do this because we are perceived as self-serving individuals who have only the corporate net gain as our goals. Those of you in academia and elsewhere are believable, credible individuals. I confess that while I was in academia I didn't do it. I would urge you, if you have the opportunity, to discuss this process, not the science specifically but the process that we use to uncover these facts to give people an understanding.

John Griesbach mentioned that he thought science education may sometime create additional worries on the part of the public. I think that's true if we go into detail. But I don't think it's true if we teach the process. I think it will begin to give them a greater feeling of control than they currently feel. It will make us better partners as we begin to develop these things within the society.

BERTRAM   Allow me just a moment to say a few things. This is not a summary and even less a synthesis. As one who perhaps only too recently was tempted to feel John Blaschke's sense of being overwhelmed by the magnitude of the problem, I find relief in discovering that various advances have been made. To move into a eucharistic mood, as a

Christian whom should I thank? I wonder who it is in my society to whom I owe a debt of gratitude. I know whom ultimately to thank for that, but there must be a lot of people to whom I'm indebted for all this. I suspect, Tom, that it may be not only the scientists, but as you and Jim have suggested, scientists who have also devoted an effort beyond the call of merely being educated. Another group which may be close to the top of the list are politicians. I almost bite my tongue when I say that. I notice that they're the favorite whipping boy not only for me but for some of you as well. I repeat the point I made before that some of them may even be Christians. I suspect that, if we look beneath the surface, one of the things that would distinguish those Christians is this Christian willingness to sin. They must have some prospect that the sin in which they engage is not ultimately held against them, that somehow it must be absolved, not cheaply but at great cost. If we were really serious about doing something about the environment, we would be moved automatically to encourage others who are worse off than we are. Certainly that's central to the Christian life.

As has been suggested, often, in order to get that third person served, we must go through someone in the middle who can be moved to act only by appealing to his or her selfishness, the very contrary of what we expect of ourselves. Either the polluters must find their polluting of the atmosphere so costly that they can't afford to do it or their nonpolluting so rewarding that they will find it beneficial to themselves, all in order to get the third party served. We can see how, ultimately beginning with ourselves, this is not only for ourselves. We have no better way to proceed than out of selflessness and a willingness to sacrifice. On the other hand, working in this complex society, the only way to implement that sacrificial care for the other, at least at a responsible level, is to have the courage to reward and penalize other people's self-interests who are involved in this complex society. For want of a better word, I'd call it politics. So I number them, those nameless politicians who are Christians, are among the people to whom I suppose I owe a debt of gratitude. Now for more edifying things, I turn the floor over to our spiritual leader.

BRUNGS   If I am your spiritual leader you are in more trouble than you know. The only real perk I have as the Director of ITEST is that I get the last word at these meetings. Man, however, cannot live on last words alone. Eric, you asked a couple of provocative questions last night, and I spent a lot of time last night thinking about them. Before I go into that, I would like to respond to your statement that I want

Christianity to be an urban religion. I said that I think it <u>is</u> an urban religion. Somehow the nature of Christianity is serenely above my wants. I think the wilderness experience of Christ lasting 40 days is in itself pretty good evidence of my contention. Forty days is not a long time in a life. Christ spent a lot more time in the urban areas.

The consummation of my life is Jerusalem, not Eden. Even in late Judaism, the notion of the future is not back to the Garden of Eden but forward to the New Jerusalem. That in itself does not prove anything. I think it is provocative, however, to consider that, while the Jewish scripture opens in a garden, the Christian scripture ends in a city. Is there some kind of divine urbanization process going on? Perhaps! I, too, love the wilderness. Some of the greatest moments of my life have been sitting and fishing in the mountains, watching the sun move across a continually changing mountain face, almost hoping that the trout don't bite.

I am concerned, however, with the notion that we "tamper" with the creation. We do much more than tamper. I think that word puts us and our activity down. We waste, we exploit, we rape, we ruin, yes. But in the long run do we tamper or is there some further purpose in this, one that we may not see yet? We've talked about hope in God. That is where all our hope is. It's the only place where we have hope. But what is our role in evolution? Is it perhaps some day to be guiding it? A possibility! We interact with "process" by becoming the *New Man* of the Scripture through our consciousness and our love. We not only have to try to understand the world, we have to learn to love it. There are elements in Christianity that have been, well, heretical. Spiritualism is one of them. We're not called to be angels. We never will be angels.

I don't like to use the word "steward." This dislike may be idiosyncratic or perhaps just simply semantic. I like to think that we're artisans, that there is something we can contribute to the beauty and the process of the world through our own creativity in Christ. For me the Golden Gate is more beautiful for the presence of the bridge. It is certainly not inappropriate. I think it adds to the "natural" beauty. I feel the same way about the New York Thruway in parts of the Mohawk Valley. I think that there's a beauty there that was not present before. I may be a rampant technologist, but I see us with the function of beautifying the earth. In our best moments we do it.

Your most provocative question, Eric, and one that kept me up for a while last night was what would Christ have us do if he came now. We can look at that two ways. What kind of world would he come to? Certainly not the one we have. It might be instructive to think of what the world would be now if he came now -- some 2,000 years after he in fact did come. We'd have to wipe out all the influences of Christianity and its derivatives. What would the world be like? It's what we used to call in physics a "thought experiment," something we can think about but can't carry out. The reality is he did come. But if he came now for the first time, would he come as a Galilean carpenter? What would he be like now if he came? I think we can say that he'd be Jewish. But where and when and what would he do? We haven't got the slightest clue. In reality he came at the only time he could come. In God's design he was born of a very specific woman at a very specific place and at a very specific time with a very specific lineage. It is the only time he could come and be who he in fact is. The same is true of us. None of us could have been conceived even a month earlier or a month later than we were and be who we in fact are. One of the beautiful things we've learned from science in this century is that there is only about a three day period in the history of the universe when anyone of us could have been born. That's about the life span of the egg in the Fallopian tube. Either we're totally meaningless or we are very important to God.

But what would Christ do now? I've thought about that, and the best I can say is that I simply don't know. Would he do exactly what he did before? I doubt it. My religious feeling is that he wouldn't. Obviously that's open to debate. What he did was done at the only time it could be done. Who knows, he might even come back as a New York City taxi driver. But it's just as possible he'd come as a Galilean carpenter.

We have to come to grips with what I call a Christian materiality. How, as matter, do we belong to God? Christianity, as I understand it, is not a movement. It's a covenanted community, a community and covenant with God. One of my lingering theological regrets over the last 40 years is that theologians have done almost nothing on the meaning of the church's defining the assumption of Mary into heaven. Why now? I think that the Holy Spirit is also very specific and thus not whimsical. Why in 1950 is the church led to pronounce as something we must believe that Mary was assumed into heaven? What does this Catholic dogma tell us about the resurrection of the body and also the resurrection of the creation, following St. Paul's statement that it, too, will share in our freedom as children of God? We're in danger of losing the realization of the survival and transfiguration of the entire cosmos in Christ.

I hope that in March we will do the theology of the creation with the heart as well as the head. I'm one of those terribly old fashioned people who still feels that we should pray our theology and not merely think it. I think it's something of the heart as much as, or maybe even more than, the head. Theology should bring us to our knees and not -- well, let's just leave it there.

John Blaschke took pity on me, seeing me limping around here. As he left, he gave me a card with a prayer of Cardinal Newman which, strangely applies to the world as well and to each of us:

> God has created me to do Him some definite service; he has committed some work to me which he has not committed to another. I have my mission -- I may never know it in this life, but I shall be told it in the next.
>
> I am a link in a chain, a bond of connection between persons. He has not created me for naught. I shall do good, I shall do his work. I shall be an angel of peace, a preacher of truth in my own place while not intending it -- if I do but keep his commandments.
>
> Therefore, I will trust him, whatever, wherever I am. I can never be thrown away. If I am in sickness, my sickness may serve him; in perplexity, my perplexity may serve him; if I am in sorrow, my sorrow may serve him. He does nothing in vain. He knows what he is about. He may take away my friends, he may throw me among strangers, he may make me feel desolate, make my spirit sink, hide my future from me -- still he knows what he is about.

That's our hope -- he knows what he is about. I want to thank especially our essayists who have given us so much to think about. None of us can doubt their knowledge and their commitment both to the world and to God. I'm always amazed by your patience with each other and with me and your charity. In conclusion, thank you and Godspeed on your return home.

PARTICIPANTS

Professor Benjamin F. Abell
Parks College of St. Louis Univ.
Cahokia, Illinois 62206

Mr. Dan Bennett
8460 Watson Road, Suite 217
St. Louis, Missouri 63119

Prof. Robert W. Bertram
7039 Westmoreland
St. Louis, Missouri 63130

John A. Blaschke M.D.
1111 N. Dewey
Oklahoma City, Oklahoma 73103

Dr. David Byers
3211 Fourth Street N.E.
Washington, D. C. 20017

Mr. Lee Carter
622 Belson Court
Kirkwood, Missouri 63122

Dr. Robert J. Collier
6926 Delmar Blvd.
St. Louis, Missouri 63130

Dr. John F. Cross
221 North Grand
St. Louis, Missouri 63103

Dr. Evelyn Crump
7519 Sheridan Rd.
Kenosha, Wisconsin 53143

Rev. Charles L. Currie S.J.
Woodstock Theol Ctr/Georgetown
Washington, D.C. 20057

Dr. Armgard Everett
2 Ridgeline
St. Louis, Missouri 63122

Dr. Charles E. Ford
221 North Grand Blvd.
St. Louis, Missouri 63103

Dr. John Griesbach
3700 Lindell Blvd. SLU Law School
St. Louis, Missouri 63108

Dr. James F. Kane
1845 Walnutway Drive
St. Louis, Missouri 63146

Dr. Eric T. Karlstrom
9730 O'Neil Court
Jamestown, California 95327

Ms Nancy Kete
401 M St. S.W. ANR-443
Washington, D.C. 20460

Prof. Jean Robert Leguey-Feilleux
221 North Grand
St. Louis, Missouri 63103

Fr. Fred Mc Leod SJ
3601 Lindell Blvd
St. Louis, Missouri 63108

Judge Thad Niemira
4924 Sutherland
St. Louis, Missouri 63109

Mr. (Thomas) & Mrs. (Rose) Quinn
3108 Savoy Drive
Fairfax, Virginia 22031-1019

Sister Maxyne Schneider SSJ
P.O. Box 116
Hubbardston, Massachusetts 01452

Dr. Thomas P. Sheahen
18708 Woodway Drive
Derwood, Maryland 20855

Mrs. Marie C. Sherman
7602 Weil
St. Louis, Missouri 63119

Ernest G. Spittler SJ
John Carroll University
University Heights, Ohio 44118

Henry N. Wellman M.D.
926 W. Michigan IUMC-UH-P-16
Indianapolis, Indiana 46223

Mr. William Witherspoon
6401 Ellenwood
St. Louis, Missouri 63105-2228

ITEST Office

Robert Brungs, SJ

Sr. Marianne Postiglione, RSM

Sr. Rosemarie Przybylowicz, OSF

ITEST PUBLICATIONS

1.* The Inner Environment: Clinical Research, Health Care Delivery, Economics, Mar., 1990, pp. 172.
2.* Science/Technology Education in Church-Related Colleges and Universities, Oct., 1989, pp. 267.
3.* ITEST Monograph: Selected Papers 1978 - 1971, Mar., 1989, pp. 121.
4.* Decision, Oct., 1988, pp. 85.
5.* Is Democracy Possible in a High-Technology Society? Apr., 1988, pp. 158
6.* Suffering: The Meaning and Management of Pain, Oct., 1987, pp. 95.
7.* Biotechnology and Law, Apr., 1987, pp. 158
8. ITEST Monograph, Fall, 1986, pp. 104 (out of print)
9.* Brain Research/Human Consciousness, Mar., 1986, pp. 148.
10.* Space Exploration and Colonization, Oct., 1985, pp. 119
11.* Science, Technology and Economic Systems, Mar., 1985, pp. 144
12.* Contribution of Science to Christian Understanding, Oct., 1984, pp. 138
13.* Artificial Intelligence, Mar., 1984, pp. 144
14.* Science and Church, Oct., 1983, pp. 125
15. Science-Faith Conflict? Mar., 1983, pp. 125
16. The Meaning of Health, Oct., 1982, pp. 118
17.* Effects of Technological Advance on the Survival of the Nation State, Mar., 1982, pp.122
18.* Warfare in the 1990's, Oct., 1981, pp. 115
19. The Patenting of Recombinant DNA, Mar., 1981, pp. 139
20. ITEST Monograph - 1981, Jan., 1981, pp. 91
21.* Government Intervention and Regulation - II, Oct., 1980, pp. 120
22. The State of the Art, Mar., 1980, pp. 137
23.* The Technology of Social Control - II, Oct., 1979, pp. 112
24.* Government Intervention and Regulation, Mar., 1979, pp. 179
25. Some Technological & Ethical Aspects of Nuclear Power, Monograph. Frank Adams, SJ, Jan.,1979, pp.124
26. ITEST Conferences, Fall, 1978, Oct./Nov., 1978, pp. 117
27. Technology of Social Control, Mar., 1978, pp. 179
28. Fabricated Man VI - Fabricated Man and the Law, Oct., 1977, pp. 100
29.* Fabricated Man V - The Religion of Fabricated Man, Oct., 1976 & "The Limitations of Science in the Solution of Social Issues," co-sponsored by NASA, Mar., 1977, pp. 96
30. Human Sexuality, Aug., 1976, pp. 144
31. Fabricated Man IV - Freedom in a Highly Technologized Society, Apr., 1976, pp. 108